Christian Behl

Estrogen – Mystery Drug for the Brain?

The Neuroprotective Activities of the Female Sex Hormone

SpringerWienNewYork

Priv.-Doz. Dr. rer. nat. Christian Behl
MPG-Nachwuchsgruppe, Max Planck Institut für Psychiatrie, Munich, Germany

Printed in Austria

Typesetting: H. Meszarics • Satz & Layout • A-1200 Wien
Printing and binding: Druckerei Theiss, A-9400 Wolfsberg

Printed on acid-free and chlorine-free bleached paper
SPIN: 10775877

With 50 Figures

CIP-data applied for

ISBN 3-211-83539-3 Springer-Verlag Wien New York

This book is dedicated to my children

Alina and Tiana

Preface

I first became acquainted with the neuroprotective activities of the female sex hormone estrogen in 1995 when I was investigating the influence of various steroid hormones on the survival of neurons. I found that estrogen (17β-estradiol, the physiological active compound) can protect cultured neurons against cell death induced by oxidative insults. Just like the well-known antioxidant vitamin E, estrogen can function as a neuroprotective antioxidant, but in contrast to the former, estrogen is a hormone, serving many functions within the body, including the brain, where estrogen receptors are also located.

This book traces six years of work devoted to gaining an understanding of the manifold neuronal activities of estrogen: the **hormone** that plays a powerful role in the brain, acting both dependent on, and independent of, receptors; the **steroid** that interacts with cellular membranes, changing neurotransmission; the **antioxidant** that prevents oxidative nerve-cell death, and the **drug** used in estrogen replacement therapy. As such, the book represents a collection of data reflecting current understanding of the various roles estrogen plays in the brain. The aim was to bring together molecular and cellular findings with clinical knowledge, establishing a link therefore between basic and clinical science. For those less familiar with molecular and cellular experimental methods and language, basic techniques are, at certain points, briefly explained.

To write a book on such a topical subject as the neuroprotective activities of the female sex hormone estrogen is a great challenge, for typical of hot topics in science, the literature is changing on an almost daily basis. It may therefore be that certain published findings appreciated by one or the other reader are not included here. I have, however, done my utmost to cover all the published work believed to be essential to the topic, and reference is frequently given to outstanding reviews of recent date that discuss certain topics in greater detail. Nevertheless, although it was my aim to achieve a well-balanced selection, the personal element involved cannot be denied.

Scientific work involves team-work, and my thanks go to a number of people, first and foremost to past and present members of my working group, who have shared my enthusiasm for, and interest in, the role played by estrogen in neurons: *Bernd Moosmann, Dieter Manthey, Frank Lezoualc'h,*

Ayako Yamamoto, Bärbel Berning, Martina Gräf, Thomas Skutella, Mauro Sparapani, Elke Hauschildt, Stefanie Engert, Monika Schäfer, Nadja Kosubek, Andreas Klostermann, Stefanie Heck, Elisabeth Güll and Sharon Goodenough. I am also grateful to the following people for collaborations, discussions, advice and support over the last years: *Florian Holsboer, Beat Lutz, Christian Krieg, Helmut Vedder, Alexander Baethmann, Andreas Beck, Matthias M. Weber, Wolfgang Burgmair, Helmut Roth, Benjamin Wolozin, Dave Schubert and Michael Jürgs.*

Scientific work also needs funding, and my sincere thanks go to the following organizations and foundations without whose support my group's work would not have been possible: Max Planck Society, the German Research Association (Deutsche Forschungsgemeinschaft, DFG), the European Union, Bayer AG, Deutsche Hirnliga e.V., Alzheimer Forschungsinitiative e.V., Deutscher Stifter Verband (Woort-Stiftung and Heller-Stiftung), Firma HERMES, Stiftung VERUM.

Last but not least I would like to thank *Helga Rüster* for all the support given.

Munich, November 2000 *Christian Behl*

Contents

General abbreviations XIII

1. Introduction 1

Estrogen – more than "just" a female sex hormone 1
Estrogen is a hormone: but what are hormones, anyway? 2
A little history 4
In the beginning there was a physiological function 4
... then there was the chemical structure 5
... an estrogen *binding-protein* 6
... and finally there was the cloning of the first estrogen receptor 7
How do hormones act? 9
Hormones 10
Estrogen – THE sex hormone and more 13

2. Estrogen is a steroid 16

Production of sex hormones in the gonadal glands 16
Biosynthesis of sex hormones 16
Hormonal changes during the female puberty and the menstrual cycle 20
Transport of estradiol in the bloodstream and catabolism 21

3. Estrogen acts via receptors 25

"Estrogen's classics" – the genomic pathway of estrogen action 25
Steroid receptors have a complex protein structure 28
Another level of complexity: estrogen acts via two ERs (ERα and ERβ) 30
ERα and ERβ: a basic comparison 34
The structural domains of ERα and ERβ 36
Modulation of the estrogen receptor function 38
Interaction of the ER with co-activators and co-repressors: of RIPs, RAPs and DRIPs 44

Selective estrogen receptor modulators – SERMs 47
To be or not to be? Are there membrane ERs? 49

4. "Non-classical" activities of estrogen 52
Rapid non-genomic effects compared to slow genomic effects of steroid hormones: what makes the difference? 52
Rapid effects of estrogen 53
Structure-dependent effects: estradiol as antioxidant 54
Reactive oxygen species (ROS): normal byproducts of life under oxygen 54
Antioxidant defense lines of the cell 56
Estradiol is an antioxidant similar to α-tocopherol (vitamin E) . 57
Dietary phenols and the blood-brain-barrier 59

5. General physiological activities of estrogen 62
Lessons from the ERKO-mice 63

6. Estrogen's actions in the brain 67
Estrogen receptors in the brain 67
Neuroactivities of estrogens in brain areas outside the hypothalamus: the effect of sex differences 68
Effects of estrogen on the cholinergic system 69
Effects of estrogen on the serotonergic and catecholaminergic system 72
Activities of estrogen on glial cells 74
Are there gender differences in brain function? 76
Sexual differentiation and gross gender differences in brain structure and function 76
Sex differences in the function of the hippocampus 77
"Non-classical" activities of estrogen in the brain 79
Estrogen as "neuroactive steroid" and estradiol's non-genomic effects at neuronal membranes 79
Estrogen's "cross-talks" with the intracellular signaling in neurons 84
Estrogen and MAP kinase signaling 85
Estrogen's "cross-talk" with other signal pathways in neurons 89
Estrogen is a phenolic antioxidant 91
The pathogenetic role of oxidative stress in the CNS 91
The brain is particularly sensitive to oxidative stress 92

7. Protection of the brain by estrogen 95
Estrogen as drug for the brain? 95

Menopause, and estrogen replacement therapy (ERT) 96
ERT for age-related degenerative diseases: general remarks 99
Estrogen and human diseases: general beneficial effects of estrogen 101
Estrogen and arteriosclerosis 101
Estrogen and osteoporosis 103
Estrogen as a drug for the treatment and prevention of brain diseases? 105
Effects of estrogen on cognition 105
Estrogen in neuropsychiatric disorders 109
Neurodegenerative disorders – Alzheimer's Disease 109
What is the cause of AD? 110
Various AD-hypotheses 111
The estrogen-Alzheimer link 119
Parkinson's Disease 123
PD and estrogen 125
Novel approaches for the treatment of PD 127
Stroke 127
Stroke and estrogen 128
Schizophrenia 130
Depression 132
To replace or not to replace? ERT and breast cancer risk 136

8. Nerve cell protection by estrogen: molecular mechanisms 142
Life is difficult, at the cellular and molecular level, too............. 142
Protective effect of estrogen in cultured neuronal cells 144
Mechanisms of nerve cell death 144
The two main routes to cell death: apoptosis and necrosis 145
Executioners of apoptosis: caspases 147
How to detect apoptosis? 149
Is there apoptosis in Alzheimer's Disease? 151
Apoptosis in post-mortem AD brain tissue 151
Apoptosis and necrosis of nerve cells in culture *(in vitro)* 151
AD genetics and apoptosis 152
Investigations of estrogen's neuroprotective activities *in vitro* 155
Intracellular molecular mechanisms of neuroprotection by estrogen 161
Direct ER-dependent neuroprotection: induction of neuroprotective genes 161
"Cross-Talks" of estradiol with neuroprotective signaling 163
MAP-kinase signaling 163
Phosphatidylinositol 3 (PI3)-kinase 167

NF-κB ... 168
NF-κB is an oxidative stress-responsive transcription factor ... 169
NF-κB has anti-apoptotic and pro-apoptotic activities ... 171
Cyclic AMP/CREB-signaling ... 172
Intracellular Ca^{2+} levels and electrophysiology ... 173
Estradiol is a neuroprotective antioxidant: ER-independent effects. 174
Oxidative stress as a general trigger of nerve cell death ... 175
Oxidative stress in neurodegeneration ... 175
Estradiol is a neuroprotective antioxidant ... 176

9. Outlook ... 182
Estrogen's neuroprotective target genes and estrogen's effects on neuronal stem cells ... 182
Following the neuroprotective trace of estrogen ... 182
DNA-array/gene chip technology and *expression profiling* ... 183
Screening DNA-arrays/gene chips to identify estrogen target genes ... 184
Estrogen and stem cells ... 186
What are stem cells? ... 186
Stem cells in the brain – the concept of mammalian neural stem cells and estrogen's effects on neural stem cells ... 187
Estrogen modulates neural stem cell generation ... 189
Final remarks: neuroprotection by estrogen ... 190

10. References ... 192

Subject index ... 225

General abbreviations

Aβ	amyloid β protein
ACh	acetylcholine
ACTH	adenocorticotropin hormone
AD	Alzheimer's Disease
AF-1	activation function-1
AF-2	activation function-2
ALS	amyotrophic lateral sclerosis
AMPA	amino-hydro-methyl-isoazol-propion acid
APP	amyloid β precursor protein
ATG	adenine thymine guanine
ATP	adenosine triphosphate
AVP	arginine vasopressin
αERKO	alpha estrogen receptor knockout
BACE	β amyloid cleaving enzyme
BBB	blood brain barrier
Bcl	B-cell lymphoma
BDNF	brain-derived neurotrophic factor
βERKO	beta estrogen receptor knockout
BrDU	bromodeoxyuridine
Ca^{2+}	calcium ion
CAD	caspase-activated deoxyribonuclease
cAMP	cyclic adenosine monophosphate
CAT	chloramphenicol acetyl transferase
CBP	cAMP binding protein
CB1	cannabinoid receptor 1
cDNA	complementary deoxyribonucleotide acid
CEE	conjugated equine estrogen
ChAT	choline acetyltransferase
CHO	chinese hamster ovary
CNS	central nervous system
COX	cyclooxygenase
CREB	cyclic adenosine monophosphate response element binding protein
CREs	cAMP response elements
CRH	corticotropin-releasing hormone
CRH-R1	corticotropin-releasing hormone receptor type 1
DAG	diacylglycerol
DATATOP	deprenyl and tocopherol antioxidative therapy of parkinsonism
DBD	DNA binding domain

DG	dentate gyrus
DHEA	dehydroepiandrosterone
DHEA-S	dehydroepiandrosterone sulfate
DHDOC	dihydrodeoxycorticosterone
DHP	dihydroprogesterone
DNA	deoxyribonucleic acid
DRIPs	vitamin D receptor interacting proteins
E2	17β-estradiol
EM	electron microscopy
ER	estrogen receptor
ERs	estrogen receptors
ERα	estrogen receptor alpha
ERβ	estrogen receptor beta
ERE	estrogen response element
ERK	extracellular-regulated kinase
ERR	estrogen receptor-related receptor
ERT	estrogen replacement therapy
ES	embryonic stem cells
ESTs	expressed sequence tags
FSH	follicle-stimulating hormone
GABA	γ-amino-butyric acid
GnRH	gonadotropin releasing hormone
GREs	glucocorticoid response elements
GSH	glutathione
GSH-Px	glutathione peroxidase
HeLa	human tumor cell line
H_2O_2	hydrogen peroxide
HGP	human genome project
HPA	hypothalamus-pituitary-adrenal
HRT	hormone replacement therapy
HSPs	heat shock proteins
5-HT	serotonin
IAP	inhibitors of apoptosis
IGF-1	insulin-like growth factor-1
JNK	c-Jun-N-terminal kinase
Kd	dissociation constant
kD	kilo Dalton
LBD	ligand binding domain
LDH	lactate dehydrogenase
LDL	low density lipoprotein
LH	luteinizing hormone
LTP	long-term potentiation
MAO-B	monoaminooxidase-B

MAP	mitogen-activated protein
MAPK	mitogen-activated protein kinase
MCAO	middle cerebral artery occlusion
MCF-7	human mammary carcinoma cell line
MPP^+	1-methyl-4-phenylpyridinium ion
MPTP	1-methyl-4-phenyl-1,2,3,6-tetrahydropyridine
mRNA	messenger ribonucleic acid
MTT	3-(4,5-dimethyltiazol-2-yl)-2,5-diphenyltetrazolium bromide
NEO	neomycin
NF-κB	nuclear factor κ-B
NGF	nerve growth factor
NLS	nuclear localization sequence
NMDA	N-methyl-D-aspartate
NO	nitric oxide
NOS	nitric oxide synthase
NSAIDs	non-steroidal anti-inflammatory drugs
$O_2^{\cdot-}$	superoxide radical
$OH^{\cdot}$	hydroxyl radical
6-OHDA	6-hydroxy-dopamine
PCD	programmed cell death
PCR	polymerase chain reaction
PD	Parkinson's Disease
PI-3	phosphatidyl-inositol-3
PKA	protein kinase A
PKC	protein kinase C
PLC	phospholipase C
PS	presenilin
RAPs	receptor-associated proteins
RIA	radio-immune assay
RIPs	receptor-interacting proteins
ROS	reactive oxygen species
sAPP	soluble amyloid β protein precursor
SDS-PAGE	sodium dodecyl sulfate polyacrylamid gelelectrophoresis
SERMs	selective estrogen receptor modulators
SOD	superoxide dismutase
TF	transcription factor
THDOC	tetrahydrodeoxycorticosterone
THP	tetrahydroprogesterone
TMP	2,4,6-trimethylphenol
TUNEL	terminal deoxynucleotide transferase-mediated dUTP nick-end-labeling

"... Methinks I should know you and should know this man;
Yet I am doubtful: for I am mainly ignorant
What place this is, and all the skill I have
Remembers not these garments:
Nor I know not
Where I did lodge last night ..."
King Lear, W. Shakespeare, Act IV, Scene VII

1. Introduction

Estrogen – more than "just" a female sex hormone

Knowledge about the female sex hormone estrogen has dramatically increased over the last two decades and, most importantly, many novel findings have proven that estrogen is much more than "just" a sex hormone. While estrogen was initially investigated almost exclusively with respect to the role it plays during sexual differentiation and maturation, it is well-known too, today for the beneficial role it plays in the nerve cells and the central nervous system (CNS). The activity of the sex hormone estrogen – and this applies to some extent to the male sex hormone testosterone, too – is neither restricted to the sex organs nor to sex function and sex behavior.

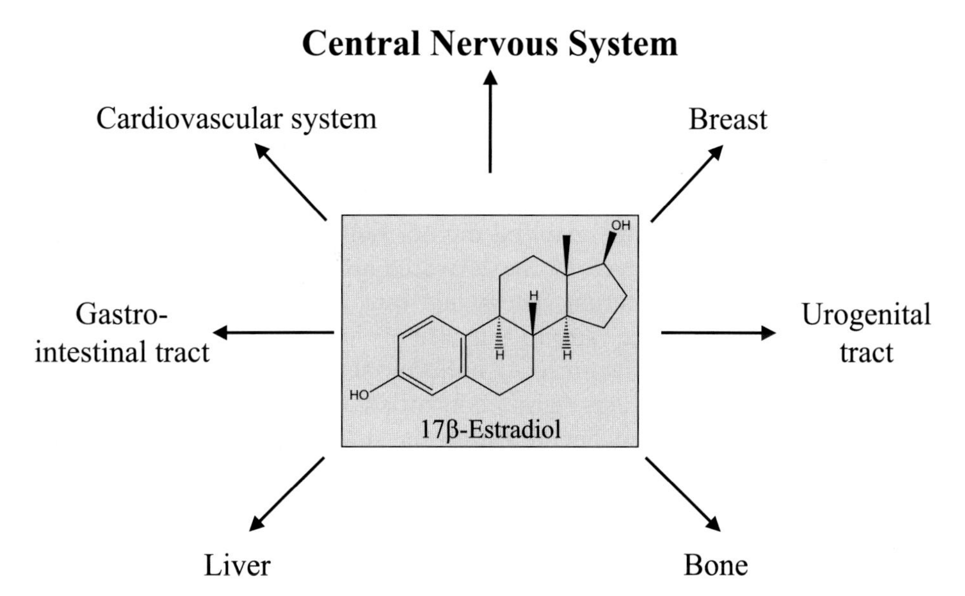

Fig. 1. Estrogen acts in various tissues throughout the human body.
17β-estradiol has the highest physiological estrogen activity compared to the other members of the estrogen family estriol and estrogen. Estradiol acts throughout the human body. Besides the traditional tissues of estrogen action such as the breast and the urogenital tract, estradiol acts also in the central nervous system

Estrogen has pivotal activities in the arteries, the bone system, the CNS, and the brain (Fig. 1). This sex hormone affects the daily functioning of our nerve cells. It affects structure and function of certain areas of the brain, including those that are of central importance for cognitive functions and memory. Interestingly, estrogen is also produced in the male body throughout life.

Estrogen biosynthesis stops with the female menopause while testosterone production in man never stops entirely. How does the female body cope with the sudden drop and later on with the lack of such a central hormone? What are the consequences for the development of age-related diseases? What is the impact of a lack in estrogen on the nervous system and the brain? One goal of this book is to increase the awareness that estrogen is more than just the female sex hormone, but an actor in many parts of the body and a major player in the brain. Estrogen is a *neurohormone*. Another goal of this presentation is to outline estrogen's protective activities in the CNS and their significance for the prevention and treatment of certain neurodegenerative disorders and, perhaps for a future drug design.

Estrogen is a hormone: but what are hormones, anyway?

The term hormone is derived from the Greek verb ωρμειν ("horman") meaning **to excite** and hormones comprise a variety of biologically active molecules. In the scientific literature the term "hormone" was first used in 1902 to describe the activity of *secretin*, which is produced by the duodenum and induces the secretion of the pancreas juice (Bayliss and Starling, 1902). Hormones are by definition *signaling molecules* produced by specific endocrine glands of the body and then secreted into the bloodstream. At certain sites in the body, hormones act on target cells. The target cell carries specific hormone receptors enabling the hormones to exert their function inside the target cell. Therefore, the activated hormone receptors transport the extracellular hormone-intrinsic signal into the intracellular compartment and – as we will see later in the case of the steroid receptors in general – into the nucleus. The presence of hormone receptors at certain target cells at one particular time is tightly controlled, and the set of hormone receptors expressed is largely dependent on the stage of development and differentiation, age and nutritional status of the organism. Hormones are produced and the secretion of hormones is controlled by the endocrine glands.

The main endocrine glands in the human body are

- the pineal gland,
- the hypothalamus,
- the pituitary gland,
- the thyroid,
- the pancreas,

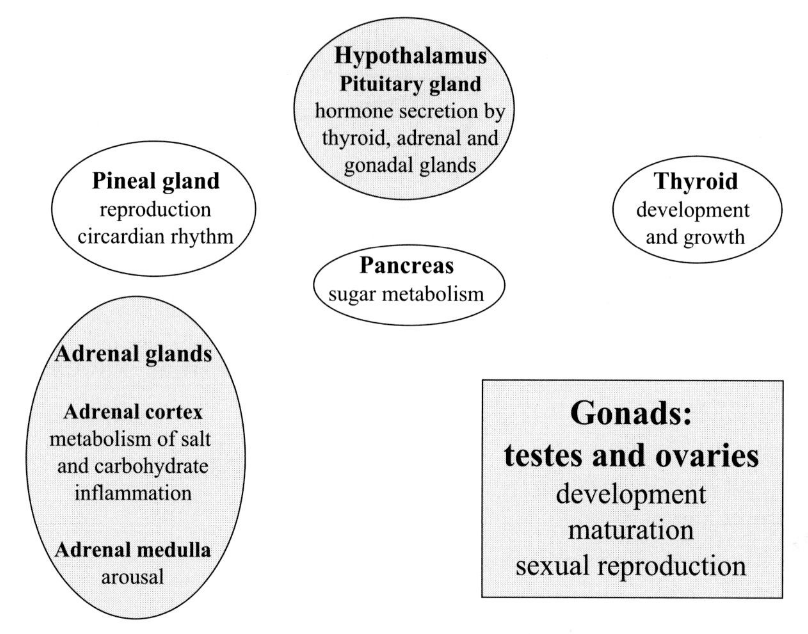

Fig. 2. The main endocrine glands and their physiological function.
The main endocrine glands in the human body and some central processes and mechanisms in which these glands are involved. The hypothalamus, the adrenal glands, and mainly the gonads will be frequently referred to in the following discussion (adapted from Baulieu and Kelly, 1990; Rosenzweig et al., 1999)

- the adrenal glands, and
- the gonads.

The anatomy of the adrenal gland consists of a cortex and a medulla. The male gonads are the testes, the female the ovaries. Each gland plays certain regulatory roles in the body (Fig. 2).

Moreover, certain endocrine glands are functionally connected and build up functional entities, which are controlled by so-called feedback and feed-forward mechanisms. Therefore, hormones link the single endocrine centers in the body and establish complex endocrine networks.

In vertebrates, hormones can be classified by their chemical structure into three major classes of principle hormones (Fig. 3):

(1) **protein hormones**, composed of amino acid chains, including the Follicle-stimulating hormone (FSH), the Luteinizing hormone (LH), the Gonadotropin-Releasing-Hormone, the Corticotropin-Releasing-Hormone, Insulin.

(2) **amine hormones**, which are composed of a single amino acid such as thyroxine and include epinephrine (adrenaline), nor-epinephrine, thyroid hormones, melatonin.

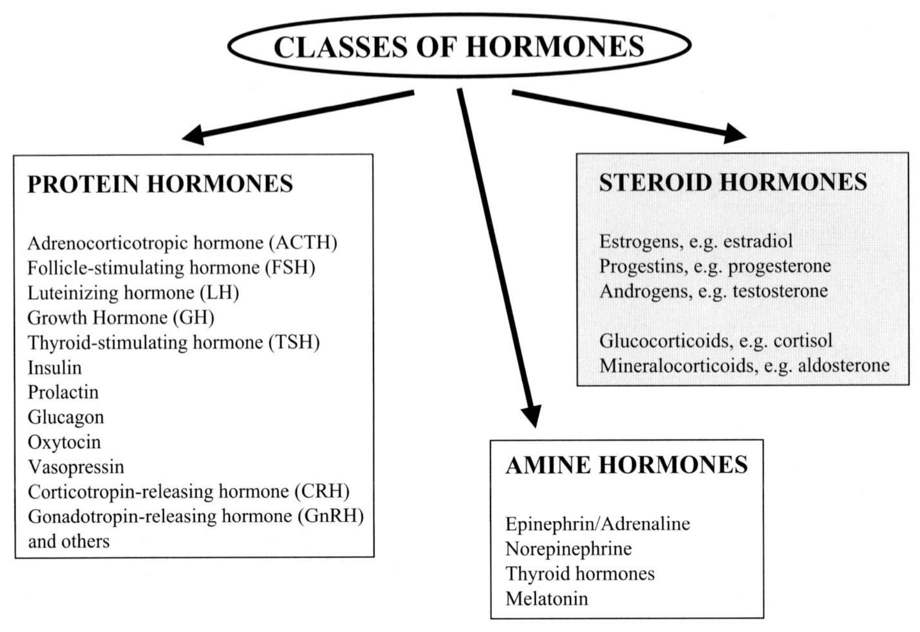

Fig. 3. Three major classes of hormones

(3) **steroid hormones**, which are composed of four interconnected rings of carbon atoms. The different types of steroid hormones vary in the number and kind of atoms attached to the basic ring structure. Important steroid hormones are the adrenal steroid hormones, glucocorticoids (e.g. cortisol) and mineralocorticoids (e.g. aldosterone), and the gonadal steroid hormones such as progestins (e.g. progesterone), androgens (e.g. testosterone) and estrogens (e.g. estradiol, 17β-estradiol).

The basic chemical structure of steroid hormones gives these molecules a rather lipophilic character and, therefore, enables them to readily cross cellular membranes, which are composed of a lipid bilayer. In stark contrast, protein hormones, which build up charged globular structures are not able to enter the cell via simple diffusion. In addition, these two classes of hormones (steroids versus proteins) interact with two different types of cellular hormone receptors and also employ different mechanisms that mediate the hormone signal via their receptor. So these receptors are quite distinct in their structural and functional characteristics (see Fig. 4).

A little history

In the beginning there was a physiological function ...

The understanding of the structure of various hormones and their function in the body developed in different stages. The Greek philosopher

Aristoteles described the effects of the removal of the male gonads in birds. With this early castration experiment he found out that the testes are of central importance for reproduction and the sexual phenotype of the birds. So the removal of an organ, here the testes, led to a loss-of-function, in this case to a loss of reproductive capability.

The experiments of the German psychologist *Arnold Adolph Berthold* in 1849 can be seen as the first systematic endocrinological research approach. He reported that young castrated roosters exert a decline in sexual behavior and secondary sexual features. By means of a replacement experiment, whereby one testis was put back into the rooster's body cavity, the typical sex behavior and sex phenotype of young roosters were restored. Furthermore, Berthold concluded that the testis releases a chemical compound into the bloodstream that changed the male behavior and male body characteristics. We know today that this chemical compound is the male sex hormone testosterone. The lacking function was restored by replacement of the hormone-producing gland.

In the nineteenth century the French physiologist *Claude Bernard* introduced endocrinology as a scientific discipline. By a variety of experimental approaches he concluded that the entire set of hormones present in the body maintains the "internal milieu" (Bernard, 1865) which builds an equilibrium of central importance for the survival of the whole organism. Later on, the principle of the body's homeostasis was created, which describes the maintenance of a constant environment in the body. As later stated, this homeostasis is built up and controlled by the endocrine glands and hormones (Cannon, 1929).

... then there was the chemical structure ...

The twentieth century witnessed novel biochemical methods for the purification of body fluids and other technical advances that paved the way for the structural identification of different hormones. The important step from the physiology to the structure was made. The synthesis of the hypothalamic hormones oxytocin and vasopressin by *Vincent du Vigneaud* was honored by the Nobel Prize for chemistry in 1955. Three years later *Frederick Sanger* received the Nobel prize for establishing the chemical structure of the protein hormone insulin. But in addition, major technical advances were necessary for the next steps to be taken in endocrinology. The development of highly sensitive techniques for the measurement of small amounts of hormones, the radioimmunoassay (RIA), a method which is still very often used, led to *Rosalyn Yalow* also being awarded a Nobel prize in 1977.

The basic hormonal functions of the female sex hormone estrogen were first realized in 1925, but its structure was not described before 1931 (for review: Corner, 1964). *Adolf Butenandt* drew up some biochemical data on estrogen in a review in 1924 entitled "Untersuchungen über das weibliche Sexualhormon" (Investigations into the female sex hormone). Applying, from today's standpoint, very basic methods of chemistry he collected a tremendous amount of information on the "progynon", as the female sex hormone was called at the time. Urine from pregnant women, prepared as an oil, was the starting point of the investigations, and he was able to crudely crystallize the hormone. He further observed that the "progynon" displayed lipid-like properties with respect to solubility, and could easily be dissolved in alcohol, acetone, chloroform and benzol but hardly in water. For the chemist, this was already vital information. He also found that the molecule structure was free of nitrogen and sulfur and a first estimate on the chemical composition was made. The chemical formula of the progynon was believed to be either $C_{23}H_{28}O_3$ or $C_{24}H_{32}O_3$, which comes very close to the actual formula of estradiol $\mathbf{C_{18}H_{24}O_2}$. It was also recognized by that time that the hormone was neither a protein nor a typical carbohydrate but displayed chemical similarities to the chemical compound family of the sterines and gallic acids. The information gathered at that time supplied the main pieces of the "estrogen-puzzle" (for review: Butenandt, 1924).

... an estrogen ***binding-protein*** *...*

There is no doubt that it was the pioneering work of *Elwood Jensen* and *Jack Gorski* that first gave insight into the possible mechanisms of estrogen's action. These scientists introduced the concept of an intracellular receptor protein for steroid hormones in general and for estrogen in particular (Jensen and Jacobson, 1962; Toft and Gorski, 1966).

These next step forward in the research on steroid hormones were made possible by various experimental tools. Indeed, one first technical breakthrough was the synthesis of radioactive-labeled estradiol, which enabled researchers to detect and follow the binding of estradiol in the tissue. This was not possible until then since the observation of a lack of physiological function *in vivo* was – more or less – the only available readout which also hinted towards possible modes of action or putative target tissues of steroid hormones. At this point the estrogen could be traced throughout the body. Imaging of estrogen was now possible. Where does estrogen go after entering the body and the bloodstream? Consequently with radiolabeled estrogen, an estrogen binding site called "estrophilin" was identified (Toft and Gorski, 1966; Jensen and DeSombre, 1973). As expected, cells in the uterus, vagina and the pituitary showed a high affinity to the labeled estrogen. It became clear that estrogen itself was the active or at least the activity-mediating compound in the target tissues and not, as thought before, an enzymatically converted estrogen which enters the tissue. This finding also stimulated the search for the estrophilin or more exactly for a hormone receptor present in certain target tissues.

A receptor for steroid hormones was initially believed to serve a simple transport function bringing the estrogen from the cytoplasm to the nucleus of the cell similar to a molecular shuttle (Jensen and DeSombre, 1973). On the one hand, the idea that the estrogen receptor is just a simple transport protein was wrong, on the other – with today's knowledge – this view is also partly right. As we will see later, estrogen-occupied receptors indeed enter the cell's nucleus, but there, the real molecular function of the receptor just starts. It is therefore in the nucleus where the main estrogen action mediated by estrogen receptors takes place. More precise knowledge about the estrogen receptor was achieved by the purification of the receptor.

Employing basic biochemical separation techniques including so-called gradient analysis, filtration and gel electrophoresis using radioactive-labeled estrogen, the estradiol-binding protein was partially purified allowing the generation of antisera (Greene et al., 1977). Now that the estrogen-binding structure could be identified by the bound radiolabeled ligand, the estrophilin, was visible. Next, monoclonal antibodies, which offered a different method for the detection of possible estrogen receptors, were generated, showing the nuclear localization of the receptor inside the cell. In addition, these antibodies initiated immunohistochemical studies of the tissue distribution of the receptor throughout the body (King and Greene, 1984; Greene et al., 1984; DeSombre et al., 1984); estrogen receptor-specific antibodies recognize epitopes of their target protein in tissue and this binding can be made visible. Moreover, it could now be observed where in the body the estrogen binding protein is expressed, even in the absence of its binding partner estrogen.

... and finally there was the cloning of the first estrogen receptor

In consequence of the availability of novel techniques of molecular biology the 1980's was the decade of the cloning of novel cDNAs coding for proteins, including several steroid receptors. The cDNA for the human estrogen receptor was first described in 1986 (Walter et al., 1985; Green et al., 1986). An immense amount of investigatory work led to the cloning of various other steroid receptors. After putting all experimental data together this work finally revealed that due to their DNA sequence homologies, all of these steroid receptors comprise a family of nuclear receptors and that there is a common mode of action of all these receptors (for review: Evans, 1988). More sophisticated methods of recombinant molecular biology led to the description and definition of the *functional domains* of the estrogen receptor. This means that the protein itself is divided into functional protein domain subunits, which serve as the basis for particular functions. On the basis of the identified domain structures, it became evident that steroid receptors, in general, are intracellular proteins that are functional transcription factors in the nucleus of the cell. This was a very important finding since it demonstrates that the receptor itself has dramatic effects on the cell's physiology by modifying gene transcription (for review: Green and Chambon, 1988). Ultimately, these findings were also the proof that the intracellular estrogen binding protein, the estrogen receptor, is not – as believed earlier – just a transport protein shuttling estrogen but rather is a functional entity and of vital importance for estrogen's actions.

As a further consequence of the molecular mapping of the estrogen receptor the existence of co-regulators and of specific agonists and antagonists was suggested (for review: Shibata et al., 1997; McKenna et al., 1999). Antagonists are molecules that have opposing effects on receptors compared to the agonist, they can block the activity of a receptor. Modulators of steroid receptor function may be of central importance, especially with respect to the manifestation of tissue-specific functions of estrogen receptors. Estrogen receptors in the brain may serve different functions than in the ovary and certainly induce the transcription of quite different genes. A further momentous step was the characterization of the crystal structure of the estrogen receptor, which indicated that upon binding of the ligand (e.g. estradiol) the conformation of the protein changes and that these changes differ significantly, depending on the binding of an agonist or an antagonist (Brozowski et al., 1997; Shiau et al., 1998). Now there was an image of the estrogen receptor, which enabled binding characteristics to be simulated, essential for the design of selective estrogen receptor modulators (SERMs). With more and more detailed information of estrogen receptor structure and function available, SERMs gain more and more importance. SERMs may also be the tool to reach tissue specificity for the stimulation or inhibition of estrogen receptors (for review: Kuiper et al., 1999). Later in the discussion this will be pointed out in more detail.

The 1990's was the decade of knock-out and gene-targeting technology. And in 1996 a second nuclear receptor for estradiol, ERβ was cloned (Kuiper et al., 1996). As good as this discovery was in explaining certain estrogen functions that could not ascribed to ERα, the modes of activity – now acting on two receptors – also made understanding of estrogen action much more complicated. The molecular disruption of a gene or the introduction of a specific mutation into a previously intact gene makes possible the investigation of the function of this particular target wild-type gene or of the pathophysiological function of the mutant gene. Several nuclear receptors have been genetically knocked-out, including the ERα, the first identified estrogen receptor, and ERβ, the second identified estrogen receptor (Lubahn et al., 1993; Krege et al., 1998). Since the mouse model with a deleted ERα-gene (αERKO) has been available now for a few years, considerable insight has been gained into the normal physiological function of estrogen and its α-receptor. Some information is already available for the βERKO mouse and much more will be learned from a combined double knock-out of ERα and ERβ (for review: Couse and Korach, 1999; Curtis and Korach, 1999). The question will be answered whether an organism can survive completely without ERs.

This brief historical summary clearly shows that knowledge about the sex hormone estrogen was achieved – as in other fields in science – stepwise, each decade having its own particular highlights. As usual in science there was always a vital interaction between scientific knowledge and the sophisticated techniques, and this interplay will further fuel further progress. The years to come, the – "decade of the human genome" – which aims to sequence and decipher the whole human genome, and the decade of the so-called *Functional Genomics*, which aims to characterize the physiological and pathophysiological functions of the identified genes, will further increase knowledge concerning the detailed and tissue-specific function of hormones (e.g. Brent et al., 1999; Watson and Akil, 1999; Hsu and Hsueh, 2000). In the very near future *expression maps* will be available that depict which estrogen receptor target genes are expressed for instance in the brain compared to the uterus or the bone. This knowledge will also be the basis for future therapies for human diseases, those associated, for example, with hormonal dysfunction. Before we take a closer look at the structure of the molecule estrogen, estrogen receptors, and into their physiological and pathophysiological functions, one has to go a few steps back and face some basic facts about hormones and their general modes of action in the body.

How do hormones act?

Hormones are messengers, which transport information from one part/tissue of the body to another. The specific information of the hormone lies in its chemistry. In general, there are three major types of "chemical communication" in the body:

- The **autocrine** function is the action of a hormone signal (e.g. growth factor) on the factor-releasing cell itself.
- In **paracrine** function the activity of a secreted signal is targeted on a nearby cell.
- The **endocrine** way of chemical communication is the release of a hormone into the bloodstream. This hormone is then selectively taken up by the target organ, which may be at quite some distance from the endocrine gland.

A few general principles of hormone action can be summarized which apply to most types of hormones, not forgetting, however, that there are always exceptions to these rules (for review: Baulieu and Kelly, 1990; Rosenzweig et al., 1999).

Hormones

- very often act in a gradual kinetic fashion, so that the hormone activity occurs long after hormone concentrations in the blood have fallen,
- do not act in a strict and simple on-and-off mode of action, but rather alter the intensity of behavior or the probability of an evoked behavior; hormones are, therefore, real physiological modulators and play a role in fine tuning,
- and behavior are frequently in a reciprocal relationship, meaning that hormones alter behavior and behavior alters hormone levels,
- have multiple effects on various tissues depending mainly on the specific expression pattern of hormone receptors,
- are generated in rather small amounts and are often released in a "pulsatile" fashion,
- are released rhythmically throughout the day; some endocrine functions have a circadian rhythm, too,
- modulate long-term metabolism in many body cells,
- directly interact with each other and complex networks are established by the "cross-talk" of various hormones,
- act via specific hormone receptors, which are genetically conserved in vertebrates.

It will be pointed out later that besides their classical mode of action through steroid hormone receptors, steroids may also act via the interaction with membrane-bound proteins and intracellular mechanisms or just by the chemical structure of the hormone itself and through direct interaction with the cellular membrane itself (receptor-independent effects). Nevertheless, the three major classes of hormones (proteins, amines, steroids) exert their well described main influence by binding to hormone receptor molecules. Two types of hormone receptors can be differentiated:

- First, protein hormones (and also most amine hormones) act via **cell surface receptors** located on or in the membrane and therefore at the periphery of the hormone target cell. Inside the cell, second messenger molecules are required to transduce the hormone signal.
- Second, due to their chemical structure, steroid hormones readily pass through the membrane and bind to **intracellular steroid hormone receptors**

These two principal hormone receptor systems also determine the time window, wherein the hormone activity is triggered: protein (and amine)

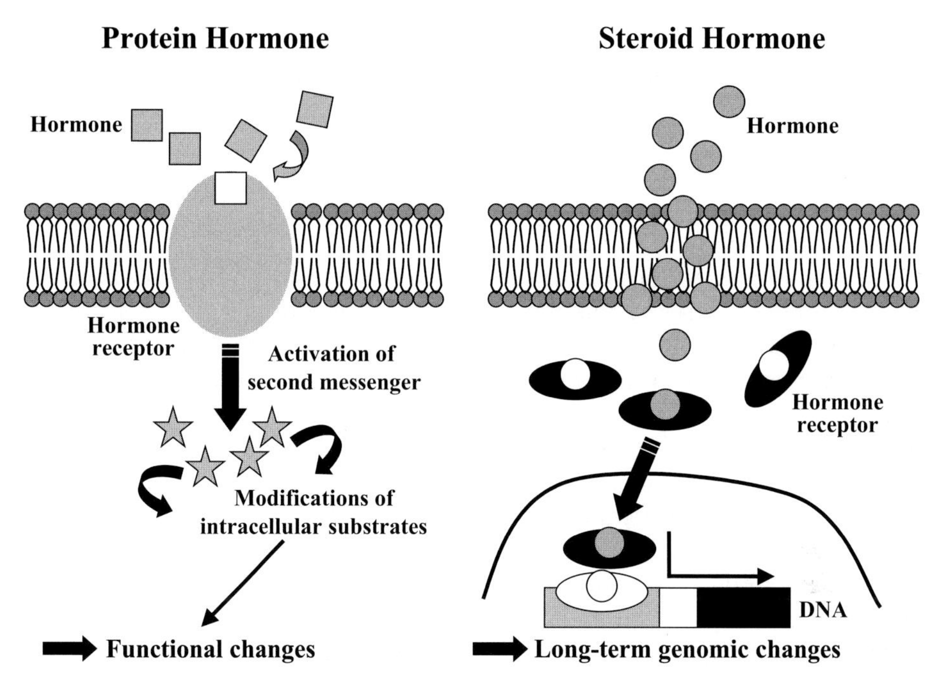

Fig. 4. Comparison of receptors for protein hormones and for steroid hormones. *Protein hormones* bind to receptors that are frequently transmembrane proteins. Inside the cells second messenger systems are activated (e.g. phosphorylating enzymes, kinases) which modify intracellular substrates. Ultimately, this chain of events leads to functional changes in the cell. *Steroid hormones* bind to receptors inside the cell. These receptors are transcription factors and translocate to the nucleus. Ultimately, steroid hormone receptors directly modulate gene transcription of the cell

hormones act rapidly, and following the binding of the hormone to its membrane-associated receptor, the intracellular part of the receptor is altered, and can activate a second messenger system. Various intracellular second messengers are known. But many signals from peptide hormones are processed inside the cell via cyclic adenosine monophosphate (cAMP) either directly or via the activation of certain so-called G-proteins. The hormone causes a change in the level of cAMP in the cell. It takes only seconds to minutes for protein hormones to activate their specific receptors, which then activate second messenger systems. Of course, cAMP is a very general second messenger but the specificity of a hormonal signal is mediated by the tissue- or cell-selective expression of a particular hormone receptor.

On the other hand, due to the fact that steroid hormones act by modulating gene transcription, and gene transcription may take some time, steroid hormones are generally considered to act rather slowly, namely within hours. The steroid-receptor complex takes action in the nucleus and steroid hormone receptors are also considered *nuclear receptors.* Estrogen belongs to the family of steroid hormones and other members of this family are the male sex hormones, also called androgens (e.g. testosterone), and the stress hormones glucocorticoids and mineralocorticoids, which are mainly produced in the adrenal glands. Estrogen receptors (both variants ERα and ERβ) belong to a protein family that comprises over 150 members and all share not only a highly conserved protein structure but also common modes of action (for review: Evans, 1988; Ribeiro et al., 1995; Warner et al., 1999). The mode of estrogen action via ERs that act as transcription factor is called the "classical" estrogen activity. Later in the discussion this mechanism will be described in more detail. Moreover, it will be discussed that today this general model is challenged and by elucidating some non-classical (receptor-independent) activities of estrogen it will be come clear that steroid hormones, in general, and estrogen, in particular, act via multiple and quite different mechanisms. This activity repertoire enables steroids to act also non-genomically and without delay.

As we have already heard, hormones affect in many ways the development, structure and function of organs in the body as well as human behavior. The interaction of different hormones controls the overall homeostasis in the body, e.g. sex hormones control the maturation of human gametes and the sex behavior or insulin regulates the overall blood glucose level. Since hormones control the central physiology in the body, it can be estimated that pathological changes in endocrine functions can have dramatic effects on survival and on human behavior. As mentioned in the introduction, the removal of the testes in young roosters obviously changed their normal sexual behavior. Further, hormones can also affect learning and memory, cognitive functions that are of central importance for mammals. For instance, it is well known that thyroid hormones are important players in the development of the nervous system. Animal studies showed that the pharmacological block of thyroid function results in a decreased formation of synapses in the brain cortex and, consequently, in disturbed learning capabilities. Hormones of the hypothalamus, including adenocorticotropin hormone (ACTH), arginine vasopressin (AVP), and oxytocin also alter cognitive functions, learning and memory. So hormones in general affect brain structure and function.

Many behavioral and neuropsychiatric human disorders are caused by **excessive** or **deficient** hormone secretion. Given the central regulatory competence and importance of hormones, in general, this can be easily understood. Interestingly, many psychiatric disorders are basically hormonal disorders. An excessive release of glucocorticoids by the adrenal glands leads to Cushing's syndrome, which is characterized by various changes in the body, such as an unusual distribution of body hair, and in behavior, such as depression (for review: Newell-Price et al., 1998).

As we will see later, the deficiency of sex hormones, especially of estrogens in the female menopause leads to a variety of physiological changes. Changes in mood can be observed during estrogen level changes such as manifested in the post-partum psychosis and during the estrous cycle. Moreover, estrogen modulates brain structure and function, and various gender differences in brain function are well known. Estrogen has a far-reaching effect on the female brain and body.

Estrogen – THE sex hormone and more

No other sex hormone has provoked as much scientific interest and publicity as the female sex hormone estrogen. Novel findings on estrogen activities in the body still cause excitement, and not just in the scientific community but also in the media and general public. Estrogen has featured on the front pages of several international magazines and newspapers and generally been a favored topic of the media. But why? What is so special about estrogen? So many hormones play pivotal roles in development, differentiation, homeostastis, and survival. For instance, if the body is not able to produce the protein hormone insulin, this insulin deficiency leads to the development of diabetes mellitus. Of course, diabetes is a major health issue and novel preventive and therapeutic approaches are well acknowledged. Still estrogen is somewhat different.

Several reasons may explain the special interest in estrogen but one stands out, namely that the menopause, with its drop in the estrogen level, constitutes a major stage in life of a female. Many women, too, take estrogens and estrogen-derivatives for contraception or as estrogen replacement therapy after menopause. In the U.S., estrogenic drugs are among the most frequently prescribed compounds. As will be pointed out later estrogen deficiency as a consequence of the female menopause can have many physiological and psychological consequences. Treatment of the above-mentioned diabetes mellitus consists of the missing hormone insulin being replaced e.g. by daily injections of insulin. With respect to the menopause-associated estrogen deficiency, one approach is to replace estrogen (estrogen replacement therapy, ERT) in the human body in order to compensate for the age-related decrease of this hormone and its possible dramatic consequences.

Other reasons for estrogen's popularity may lie in the public awareness that estrogen has a fundamental role to play during sex maturation, in sex differentiation, and in sex behavior, and that the menopause also coincides with the onset of a variety of age-related diseases that may affect large parts of the population. These include diseases of various organs and tissues of the human body, such as skeletal bone (osteoporosis), the blood circulation system (arteriosclerosis), and breast tissue (breast cancer). In addition, the central nervous system (CNS) and the changes affecting it in the case of estrogen deficiency has received major attention, especially with the increasing knowledge of the influence of estrogen on the neuronal tissues. In particular, the role of estrogen in Alzheimer's Disease (AD), the most prominent and most frequent cause of dementia, has now become the subject of much attention and the possible role of estrogen and ERT for the prevention and/or treatment of AD is currently under preclinical and clinical investigation.

The following collection and discussion of data on estrogen and ERs mainly focuses on the molecular mechanisms that mediate estrogen's *protective activity* in nervous tissue and the brain. Since many of estrogen's activities are controlled by estrogen receptors, emphasis is also put on the knowledge of the structure of ERs and their possible modes of action. It will become apparent that estrogen has target sites throughout the body besides the sex organs. The main part of the collection describes what is currently known about estrogen's neuroprotective activities focusing on basic experimental findings as well as on completed clinical studies. These are discussed mainly in the context of neurodegenerative disease such as AD. This disorder forms the center of discussion not only since it is the author's major topic of research but also because various clinical trials employing estrogen for prevention or for therapy have just recently been completed. AD is a disease that threatens every one of us and is a deadly disease for which there is no cure. Its pathogenesis is still an unsolved mystery to a larger extent. Therefore, at certain points in the book, the impact of the given information for neurodegenerative disorders such as AD will be underlined.

This book also aims to introduce some of the landmark molecular and cellular experimental systems and strategies which are the current tools for the study of cellular functions and for the discovery of novel pharmaceutical drugs. The goal here is to introduce techniques with simplified cartoons in order to make the subject matter understandable for the non-specialist in the field.

Although the data on the neuroprotective activities of the female sex hormone estrogen are very exciting and one tends to dive in headfirst, in order to understand these protective activities a detour has to be entertained. During the excursion, molecular and cellular data on estrogen and ERs are first presented in order to prepare the reader for the discussion of estrogen's role in neuroprotection or as a potential drug for the brain. In other words: to reach a mountain's summit one needs to be well-equipped, and the molecular biology and the physiology of the female sex hormone estrogen as a whole can be likened to a high and precipitous mountain.

Summary: Hormones are messengers of information and molecules that convert extracellular signals into intracellular activities. Hormones can act in an autocrine, paracrine and endocrine fashion. They are produced by endocrine glands such as the female or male gonads, which produce sex hormones, estrogens and androgens. In addition to steroid hormones, amino hormones and protein hormones also exist. Estrogen is the female sex hormone and in target tissues throughout the body estrogen acts via intracellular estrogen receptors (ERs). The two types of ERs (ERα and ERβ) are nuclear transcription factors and belong to a large superfamily of nuclear receptors as revealed by a huge cloning effort. In contrast to the family of nuclear receptors, which act at the genomic DNA, receptors for protein hormones are frequently associated with the cellular membrane. Second messenger systems transduce the hormone signal from the outside of the cell into the cell's interior. Estrogen is one of the most intensively studied steroid hormone and it is now considered not only as a sex hormone but rather as a neurohormone with a wide range of activities in the CNS.

As mentioned before, the main topic of this book is the summary of what is currently known about estrogen's neuroprotective activity. A great emphasis is put on the possible molecular mechanisms of estrogen's actions. Therefore, to understand the exciting findings concerning estrogen-mediated neuroprotection the following presentation is developed in three major steps:

- General modes of action of steroids.
- Estrogen's general activities in nervous tissue.
- Estrogen's neuroprotective activities.

2. Estrogen is a steroid

Production of sex hormones in the gonadal glands

Sex hormones are produced and secreted by the gonads. Anatomically, both the female ovaries and the male testes consist of two distinct compartments. One compartment produces the hormones and the other the gametes. The hormones produced by the gonads are not only important for the production of gametes but also for functions in target tissue distant from the gonads, e.g. the induction of reproductive and other behavior in the brain. In the testes, sperm-producing Sertoli cells and testosterone-producing Leydig cells can be differentiated. Testosterone production and release is tightly controlled by the luteinizing hormone (LH) secreted by the anterior pituitary. LH secretion is controlled by the gonadotropin releasing hormone (GnRH) of the hypothalamus.

The ovaries produce the female gametes (ova, eggs), generate and secrete hormones in a cyclic fashion. An ovarian cycle in the human female takes approximately 4 weeks. Human ovaries produce estrogens and progestins. The most prominent and important physiologically active estrogen is 17β-estradiol and the most important progestin is progesterone. The production of the ovarian hormones is regulated by two hormones from the anterior pituitary, LH and follicle-stimulating hormone (FSH), which themselves are again under the control of GnRH (Fig. 5).

Biosynthesis of sex hormones

The three classes of sex hormones (androgens, estrogens, and progestins) are chemically related. The biosynthesis of all of these steroids is linked and they are all derived from the same chemical precursor cholesterol. The biosynthesis of the sex hormones takes place in the steroid-producing glands, the testes and the ovaries. Estrogens are derived either from androstenedione via a sequence of enzymatic reactions or from the androgen testosterone through the aromatase-reaction. The latter consists of a C19 hydroxylase and a hydroxysteroid oxidoreductase. This "aromatase reaction" means conversion of a keto group at the C3 position of the steroidal

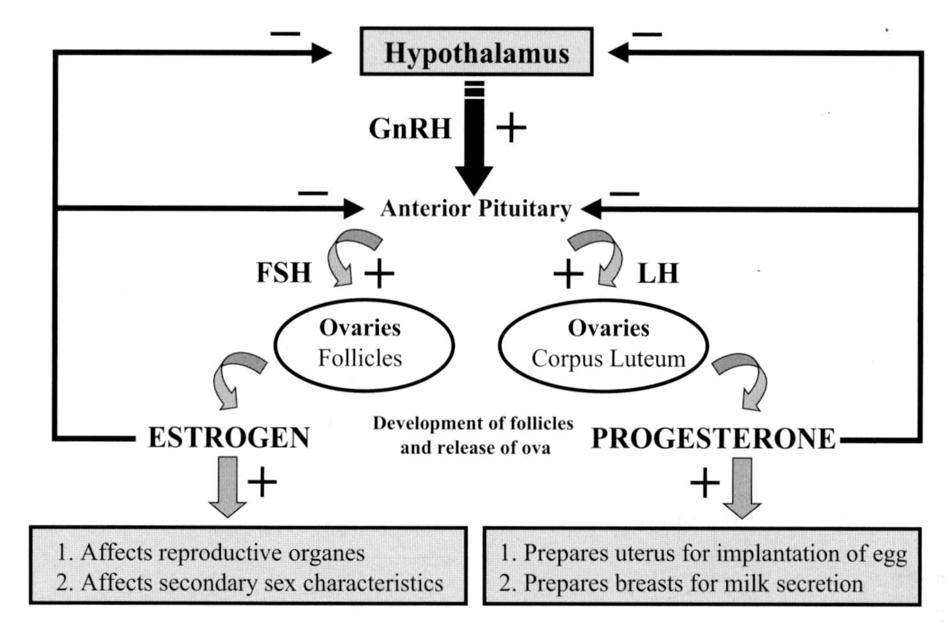

Fig. 5. Regulation of the secretion of the female sex hormones.
The hypothalamus releases the gonadotropin releasing hormone (GnRH) that acts on the anterior pituitary. The pituitary secretes follicle stimulating hormone (FSH) and the luteinizing hormone (LH), which then stimulate the secretion of estrogen and progesterone in the ovaries. The steroid hormones act in certain target tissues and negatively feed back onto the pituitary and the hypothalamus. This chain of events is frequently called Hypothalamus-Pituitary-Ovary-Uterus-Axis (adapted from Baulieu and Kelly, 1990; Rosenzweig et al., 1999)

A-ring into a hydroxyl group, which generates an aromatic ring (ring A) in the steroid molecule. This reaction fundamentally changes the physiological as well as the biochemical activities of the molecule now called estradiol. Estradiol is a phenolic molecule and belongs to the chemical family of aromatic alcohols (Fig. 6).

Estradiol is not the only secretory steroid generated. Besides estradiol human ovaries secrete various steroids including progesterone, pregnenolone, 17α-hydroxyprogesterone, dehydroepiandrosterone (DHEA), androstenedione, testosterone, and estrone. Estradiol, estriol and estrone all belong to the family of estrogens but in the literature and also in this book mostly the term "estrogen" is used synonymously for the biologically most active estradiol (17β-estradiol; E2) unless otherwise stated. The synthesis of these steroids is strictly compartimentalized meaning that although all cell types in the ovary have the potential to synthesize all these steroids, each compartment of the ovary (follicle, corpus luteum and stroma) secretes a special compound (Fig. 7).

As mentioned, the secretion of steroid hormones, testosterone and estrogen, is controlled by neuroendocrine feedback loops involving the

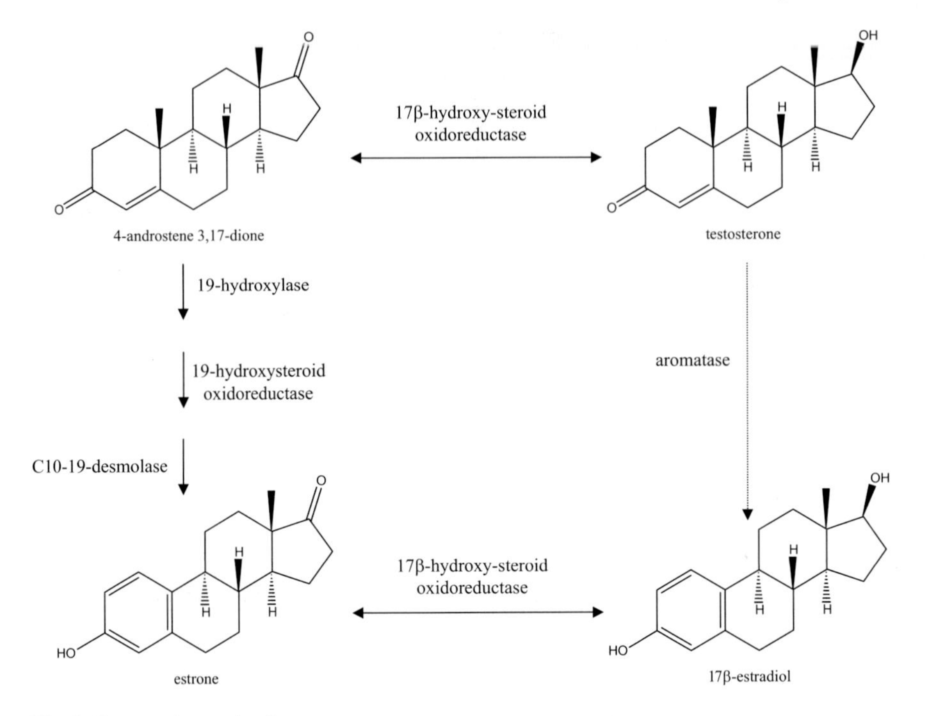

Fig. 6. Estrogen's synthesis.
Estrogens are derived from 4-androstene 3, 17-dione and testosterone (via the aromatase reaction). Estrone is derived from androstenedione. Various organs throughout the body can convert estrone into estradiol. The estrogenic activity of estradiol is around ten times greater compared to estrone

hypothalamus and the anterior pituitary. The production and secretion of the female sex hormone is pulsatile since it depends on the pulsatile secretion of gonadotropin releasing hormone (GnRH), Luteinizing hormone (LH) and follicle stimulating hormone (FSH). GnRH is secreted by the hypothalamus and induces the secretion of FSH and LH from the anterior pituitary. Finally, inside the ovaries, specifically in the follicle's granulosa cells and theca interna cells, whose activity is controlled by FSH and LH, estradiol and estrone is synthesized via the androgen steroid intermediates. This chain of events is frequently called "Hypothalamus-Pituitary-Ovary-Uterus-Axis" (Fig. 5).

Different organs contain different amounts of sex hormones and no steroid can be found exclusively in either males or females. But the proportion of these steroids differ quite dramatically. For instance, only a minor amount of testosterone is converted to estradiol in the testes while the ovaries convert almost all of the testosterone into estradiol. Interestingly, in male organisms some of the testosterone which enters the brain is locally aromatized into estradiol. The testes of men indeed secret only some

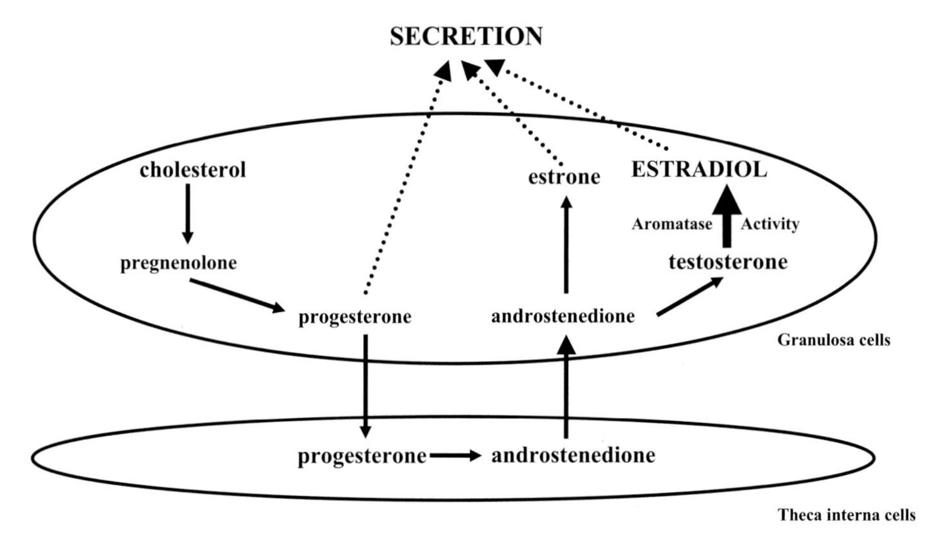

Fig. 7. Compartimentalization of estrogen's biosynthesis.
Estrogen's biosynthesis occurs in compartments. Granulosa cells of the ovary secrete estradiol, estrone, and progesterone (adapted from Michal, 1999)

estrogen but – as mentioned – testosterone can be aromatized to estradiol in various areas of the brain, including the hippocampus (Marcus and Korenman, 1976; Naftolin and Ryan, 1975). Receptors for testosterone (e.g. androgen receptor) are expressed also in the hippocampus of rats and humans (McEwen, 1980; Tohgi et al., 1995) suggesting that testosterone may also influence hippocampal function such as memory in males. Indeed, testosterone improved explicit spatial memory in treated male rodents (Flood et al., 1992). Estrogen has also major effects on the structure and function of the brain. Estrogen has a high impact on the development of the nervous system and influences brain functions throughout the adult life. Many effects of estrogen have been observed in the brain structure which is acknowledged to be involved in higher cognitive functions. Gender differences in brain structure and function as well as in disease development will be presented later.

Interestingly, elderly male individuals have higher estradiol levels than the age-matched (non-E2-supplemented) elderly women. The major source of estradiol in women are the ovaries, which become atrophic after menopause and from then on secrete only a minor amount of estradiol and estrone (Longcope, 1986). In elderly men testosterone levels also decrease with age (Tenover, 1999). Nevertheless, 80% of plasma estradiol arises from the peripheral conversion of testosterone and testosterone production never stops entirely in the elderly which makes testosterone available for the conversion to estradiol a lifelong in men (Tenover, 1999).

Hormonal changes during the female puberty and the menstrual cycle

The female puberty is marked by the induction of a menstrual cycle. The menstrual cycle itself is characterized by periodical changes in the plasma levels of FSH, LH, progesterone and estradiol (Fig. 8).

Obviously, the mechanism of this regularly occurring cycle is under a tight control of various endocrine *feed back* and *feed forward* loops. These interactions are complex enough and would need a much longer discussion for satisfactory understanding. Briefly, during the menstrual cycle estrogen induces massive physiological, morphological and structural changes of the uterus tissue, such as the proliferation of its mucosa. After ovulation, it is the progesterone that changes the physiology and pushes the endometrium into the phase of secretion. The endometrium is a specialized tissue that undergoes cyclical growth and shedding and which allows the implantation of the proliferating embryo after conception. Without conception the next step is the menstruation, in which the apoptotic follicle is repelled, a process that occurs as monthly "bleeding". Indeed, as elegantly outlined in a review in the journal *Frontiers in Neuroendocrinology* in 1998 our understanding

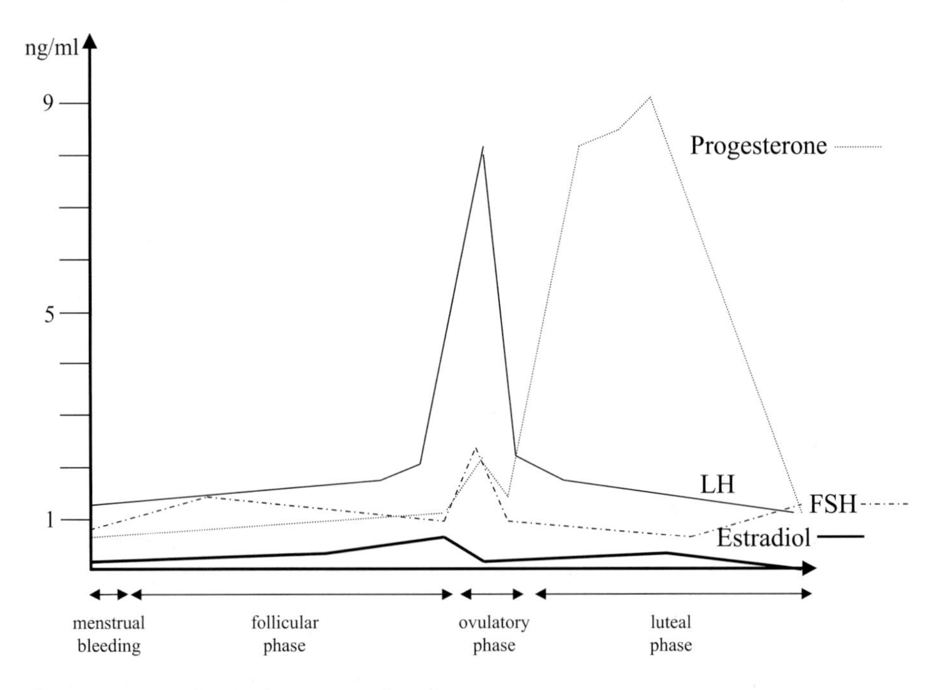

Fig. 8. Hormones during the menstrual cycle.
Changes of the level of LH, FSH, progesterone, and estradiol during the human menstrual cycle (adapted from Michal, 1999)

of physiology and regulation of the menstrual cycle has changed due to advances in experimental technologies and approaches. It is now believed that the menstrual cycle is determined mainly by the ovary itself and that there is nevertheless an intensive cross-talk between the ovaries, the pituitary and the hypothalamus. Ultimately, the goal of the menstrual cycle is to produce a single mature oocyte each month from puberty to menopause (for review: Chabbert-Buffet et al., 1998).

Interestingly, the cyclical variations of the gonadotropins are regulated by the ovarian steroids estradiol and progesterone and by the peptide inhibin. With modern molecular techniques, such as DNA cloning procedures and recombinant technologies, the mechanism of action of gonadotropins were much better characterized and understood. The complementary DNAs (cDNAs) of gonadotropins and their receptor have been cloned. Also, the cellular mechanisms of endometrial maturation, under the control of estradiol and progesterone, are better understood. Most importantly, the genes that are involved in the implantation of the embryo into the endometrium are currently identified (e.g. integrins). Last but not least, the cellular and molecular events leading to the endometrial shedding are being elucidated, especially the role of metalloproteases and angiogenic factors. The identification of the key molecular players paves the way for the development of new treatments for infertility and the design of new contraceptive techniques (for review: Chabbert-Buffet et al., 1998).

Transport of estradiol in the bloodstream and catabolism

Sex hormones act in an endocrine fashion which means that their site of action (target tissue) can be quite distant from the hormone production site. Therefore, transport of hormones to those tissues that express a particular type of steroid hormone receptor is needed. This transport is managed mainly through the blood plasma. In the blood the steroid transport is controlled and managed by the binding of the steroids to certain proteins. There is a constant equilibrium between the bound and the free fraction of estrogen, the latter being the physiologically active fraction. Various types of binding proteins are known, some of them bind certain steroids with a higher affinity than others. For instance, estradiol is bound with a 1000 times higher affinity to the so-called sex steroid-binding plasma protein than to albumin in the blood plasma (for review: Baulieu and Kelly, 1990). Technically, the binding of steroids to carrier proteins prevents the random diffusion in blood vessels and into blood cells and ensures the presence of these molecules at specific target sites. Binding of steroids and other active compounds to proteins is a basic regulatory mechanism, which takes place also inside the cells to regulate the amount of certain molecules in a particular region and to guard and protect proteins. For instance, also steroid receptors, in general, are not freely diffusing inside the cell but are bound to chaperone proteins (e.g. heat shock proteins, HSPs) that mask the steroid receptor. Following the binding of the steroid to its receptor HSPs are released from the receptor molecule. For the activation of the estrogen receptor (ER) induced by the binding of E2 to the receptor and for its translocation to the nucleus the chaperoning HSPs need to be removed from the ER-HSP complex. Until recently the power of this step in terms of regulatory function of receptor molecules was somewhat underestimated. HSPs comprise a complex family of proteins with different molecular weights, which are functionally connected. Fascinating biochemical studies are ongoing that uncover novel functionas supplied by the chaperones, which may affect ER activity (for review: Ellis and Hartl, 1999; Saibil, 2000).

Besides the biosynthesis of biologically active structures and molecules other important regulatory mechanisms are the de-activation of the compounds, their catabolism, and clearance from the body. The major catabolic principle applies to all steroidal molecules. The polycyclic carbon ring of the steroidal structure is not entirely broken up, which would need a sequence

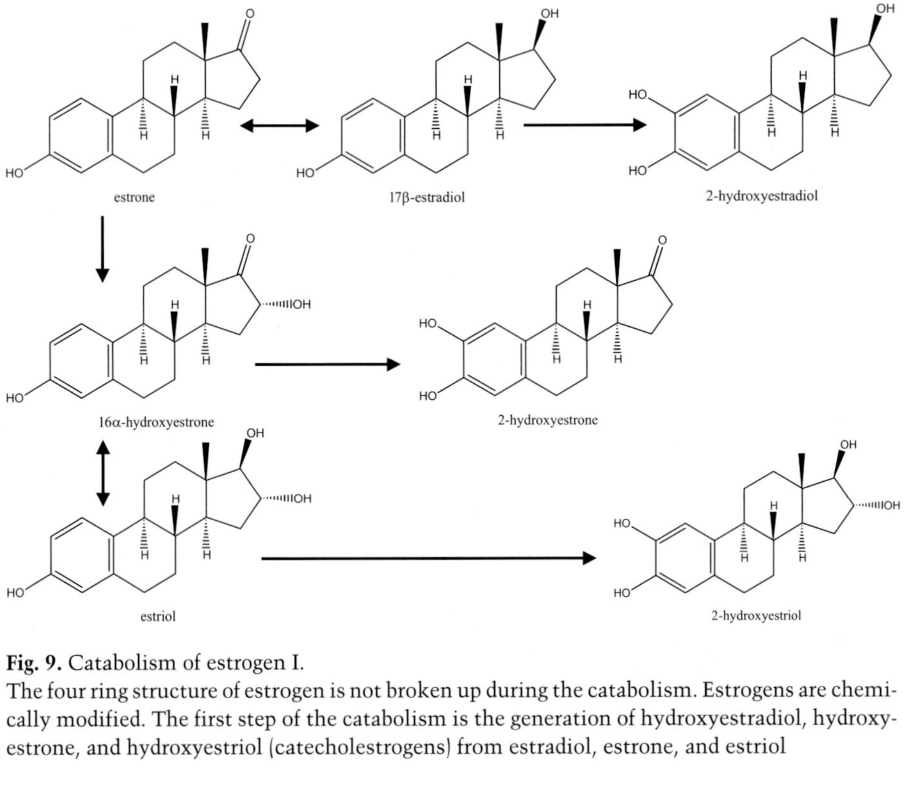

Fig. 9. Catabolism of estrogen I.
The four ring structure of estrogen is not broken up during the catabolism. Estrogens are chemically modified. The first step of the catabolism is the generation of hydroxyestradiol, hydroxyestrone, and hydroxyestriol (catecholestrogens) from estradiol, estrone, and estriol

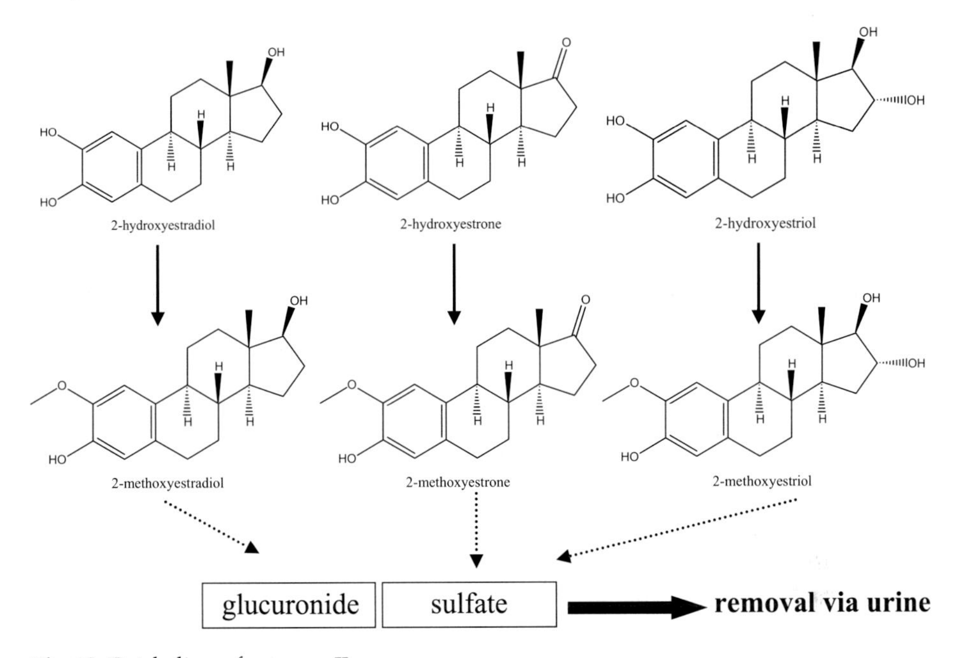

Fig. 10. Catabolism of estrogen II.
The methoxylation of the catecholestrogens is the second and the formation of glucuronides and sulfates the third step in the catabolism of estrogens. The chemical modification of the estrogen structure generates more water soluble derivatives which are, ultimately, removed via the urine

of chemical and highly energy-dependent reactions. Rather, the carbon ring is enzymatically modified, which leads to its de-activation and ultimately the elimination mostly through the urine. Chemically, the catabolism of steroids consists of a series of reduction and hydroxylation reactions and in a final step acids are conjugated to the steroids rendering these molecules water and urine soluble (Figs. 9 and 10).

Interestingly, there is a reversible equilibrium between estradiol and estrone, which may lead to certain types of enzymatic reactions depending on the particular starting point (estradiol or estrone). During the catabolism of estrogen *catechol estrogens* may be formed, which are also known to be generated in large amounts in the hypothalamus of the brain and in the pituitary (Fig. 9). Via O-methyl-transferases methoxyderivatives of the hydroxyestrogens are formed. In a final step conjugating enzymes such glucuronyltransferases and sulfotransferases catalyze the formation of glucuronides and sulfates. This last step increases the solubility of the estrogenic compounds in water (Fig. 10).

Summary: Cholesterol is the chemical precursor of the synthesis of steroids including estrogens. In the non-pregnant female, the main sites of estrogen production and secretion are the ovaries. The term "estrogens" describes the three biologically relevant estrogens estradiol, estrone, and estriol but estradiol has the highest biological activity and is, therefore, the most potent estrogen in the estrogen family. Estrogens are mainly produced in the ovaries but estradiol may be locally synthesized in the brain. The chemical conversion of testosterone to estradiol is regulated by the enzyme cytochrome P450 aromatase. The secretion of estrogens from the ovaries is highly dependent on the estrous cycle and so are the estradiol levels in the plasma. The transport of sex hormones to their target tissues is managed through the bloodstream, where the steroids are bound to high affinity binding proteins. In summary, the female ovaries are the main producers of estrogens but besides estrogen also secrete progesterone. Interestingly, also the testes secrete small amounts of estradiol. But certainly, testosterone is the main product of the testes.

3. Estrogen acts via receptors

"Estrogen's classics" – the genomic pathway of estrogen action

Since estrogen belongs to the family of steroids and estrogen receptors are part of the steroid receptor superfamily the following summary applies to steroid receptors, in general. As introduced, in contrast to protein hormones, which act via membranous hormone receptors, steroid molecules readily cross the cellular membrane because of their high lipophilicity. Although diffusion is believed to be the main mechanisms of the penetration, for certain steroids and in certain tissues a facilitated diffusion has been proposed. Early investigations have addressed the question whether estrogen's entry into cells of the uterus is not a simple diffusion but is rather a protein-mediated facilitated process (Milgrom et al., 1973). Indeed, such a facilitated or enhanced uptake process would introduce an additional level of regulation of estrogen activity since it would be the "transport-protein" that controls the entry of the steroids into the cells.

Through its passage estrogen can interact with various membrane structures such as integral ion-channels. Concerning the possible interactions of estradiol with membrane-bound receptor systems, e.g. the 5-HT_3-receptor and many others, and considering also the possibility of a specific estrogen receptor located inside or attached to the cellular membrane (Wetzel et al., 1998; for review: Rupprecht and Holsboer, 1999; Schmidt et al., 2000; Levin, 2000), these observations currently experience a revival and will be discussed later in more detail. It may be speculated that the binding of steroids to membrane structures (e.g. integral membrane receptors) can concentrate steroids at certain sites and therefore increase the diffusion across the membrane. Whether there is, indeed, a special protein that pulls estrogen through the membrane is not very likely. So, the membrane passage of a steroid molecule is mainly working through diffusion.

Once inside the cell the "classical" mode of action of steroid hormone activity, the so-called genomic pathway is activated. This process includes three main steps.

- The steroid binds to the receptor:
  Binding of the ligand (e.g. 17β-estradiol) to the receptor induces confor-

mational changes of the receptor into an "activated (estrogen) receptor". This receptor activation goes along with the dissociation of the chaperoning HSPs (hsp 70 and hsp 90). In some recent models of steroid action, such receptors are also considered to be mainly present in the nucleus. Nevertheless, at one point the receptor needs to translocate into the nucleus.

- Translocation and dimerization of the receptor:

  The receptors form dimers which may include ERα and ERβ homodimers as well as ERα and ERβ heterodimers and translocate to the nucleus. So it is not a single receptor molecule that is occupied by estradiol which is active in the nucleus but rather dimers of steroid-activated receptors.

- Receptor action at the DNA level – regulation of transcription:

  ER dimers bind to promoters of estrogen target genes that contain so called estrogen response elements (EREs) and induce transcription. The transcription process itself is highly complex and many different proteins need to find together in a protein complex before the transcription actually can start. For the activation of the transcriptional machinery, co-activator proteins and various components of the RNA polymerase II transcription

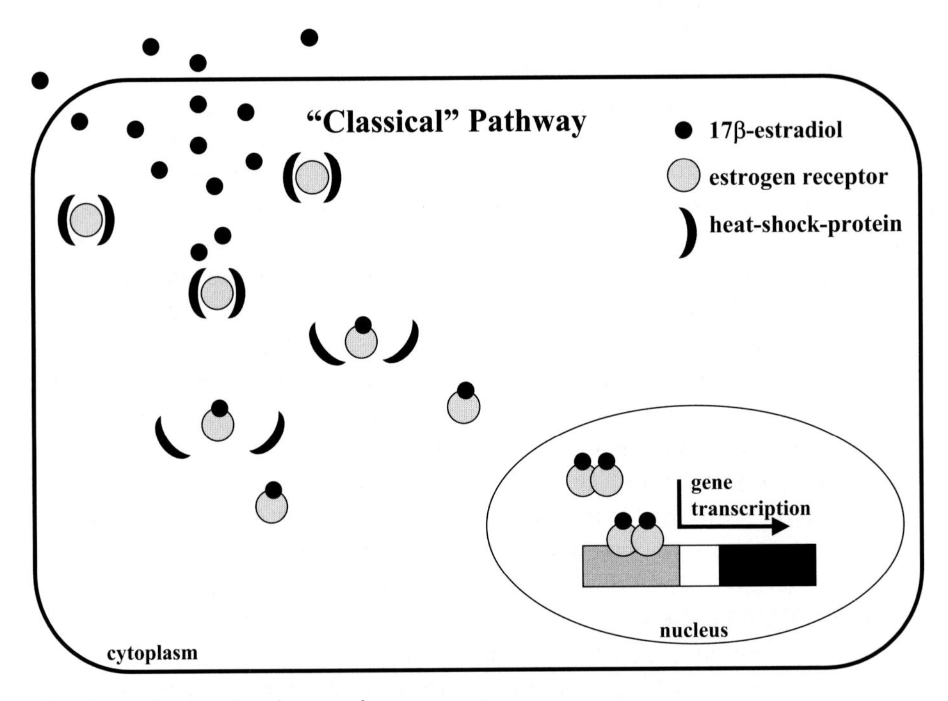

Fig. 11. The "classical pathway" of estrogen action.
The "classical pathway" of estrogen action consists of three steps. Step 1. Estradiol enters the cell and upon binding of estradiol to the receptor the heat shock proteins, which are chaperoning the ER, are released. Step 2. The estradiol-bound ER translocates to the nucleus and binds to estrogen-response-elements (EREs). Step 3. Gene transcription is induced

initiation complex are assembled together with the estrogen-liganded ER dimers at the EREs.

This "classical" three-step-model is depicted in Fig. 11 and explains that estrogen receptors bound to the cellular DNA are functional transcription factors.

A protein hormone is too big and also charged which prevents its passage through the membrane. It binds to its receptor in the membrane which is occurring – more or less – in one step. The protein hormone needs to be recognized by its receptor, which then turns on a cascade of intracellular downstream signaling events. In contrast, the steroid binds to the receptor and follows the receptor in to the nucleus. The steroid receptor on the other hand needs to be a specially designed molecule since it has to fulfill quite different functions. The ability to perform these different functions is provided by the structure of the steroid receptor. In these receptors, three major protein domains are present that build the structural prerequisites for the three steps of the "classical" steroid receptor action. Again, since ERs belong to the family of steroid receptors, also in ERs the following steroid receptors domains can be defined:

- ligand binding domain,
- modulatory domain,
- the DNA-binding domain.

In the last years it became more and more evident that the proteins that interact with ERs for instance with the modulatory domain play major roles in ER-controlled transcription. And indeed, many different proteins that promote the ER-driven transcription (co-activators) or those that soften the transcription (co-repressors) are known.

Estrogen receptors (like all steroid receptors) are functional transcription factors and can be simplistically described as "molecular switches" that turn on (or off) the transcription of specific genes. As will be pointed out later for ERs this picture of ER activity is highly simplified and it will turn out that the specific activity of the "molecular switch" highly depends on, and can be modified by, additional adaptor proteins that associate with the active ER and that may modulate tissue specificity.

Many different intracellular transcription factors are known and they differ in their sequence-specificity at the promoters of the genes and in their protein structure. But common to all transcription factors is the fundamental function to be the gene-specific anchor for the basic transcriptional machinery including most importantly the RNA polymerase II. For instance, the presence of ERs bound to DNA alone would not be sufficient to induce transcription and without the other anchor proteins and, ultimately, the fully assembled protein complex the polymerase would drift along the DNA lacking a specific site to start the transcriptional activity. In general, transcription factors transform extracellular signals into genetic signals with long-lasting effects, a basic function which is of vital importance. Some transcription factors such as the cyclic AMP response element binding protein (CREB) are specifically phosphorylated by kinases that have been directly or indirectly activated by extracellular stimuli. Others such as the steroid receptors are binding their specific ligand with high affinity and translocate – the steroid still bound – into the nucleus determined to bind to the DNA. In contrast to peptidergic hormones such as the corticotropin releasing hormone or vasopressin, which bind to transmembrane receptors and which use intracellular phosphorylation and second messengers as signal transducers, the steroid transports its intrinsic signal from the extracellular compartment directly into the nucleus and induces long-term changes in the gene expression pattern (for review: Lewin, 1974; 2000).

Today bearing in mind that the sequence of the humane genome is going to be deciphered soon, or at least mostly sequenced (which is the prerequisite to clarify the genetic information hidden in the genomic DNA) and that sophisticated molecular and also imaging techniques are available, the early experiments of Jensen and Jacobson in 1962 appear rather hand-crafted but, nevertheless, resulted in landmark findings. By injecting radioactively labeled estradiol (^{3}H-estradiol) into rats they studied the tissue distribution of the radioactivity. With these simple experiments a distinction was made between estrogen target tissues including the uterus and vagina, which accumulated and retained the radioisotope and non-target tissue, which showed only lower concentrations of radioactivity for a rather short period of time (Jensen and Jacobson, 1962). The modulators of this accumulation and retention of estrogens in cells of the target tissue have been later identified as soluble cytoplasmic estrogen receptors (as opposed to e.g. membranous peptide receptors).

With well-developed high resolution microscopical methods including electron microscopy (EM) and confocal fluorescence microscopy the translocation of steroid receptors upon steroid addition can be nicely demonstrated. But how are these receptors built? What is their basic molecular architecture?

Steroid receptors have a complex protein structure

Steroid receptors are complex oligomeric proteins consisting of several protein subunits. Steroid receptors comprise a large family including the glucocorticoid receptor, estrogen receptors and many others and one main common feature is their high hormonal specificity. Agonists and antagonists of steroid receptors bind with certain affinities, e.g. the ER binds estradiol, the main physiologically active estrogen, more strongly than estrone or estriol, the "side products" of estrogen's catabolism. Antagonists such as tamoxifen or ICI compounds (e.g. ICI 182,780; Howell et al., 2000) also have high affinities for the estrogen receptor and are potent antiestrogens. Antiestrogens are not only very useful tools to study estrogen receptor function but even more important may be used also as potential drugs in the fight against certain types of cancer, whose growth depend on the transcriptional activity of ERs (e.g. certain breast cancers). Blocking the ER with an antagonist can prevent the growth of certain types of breast cancer.

Despite many similarities between the various steroid receptors and although estrogens belong to the family of steroid molecules such as progesterone or cortisol, the main human corticosteroid, these do not cross-react with the ER. This demonstrates the strict structural requirements given by the receptor for the ligand to ensure specificity. Initially pharmacological techniques like receptor binding and *in vivo* replacement studies (e.g. removal of testes or ovaries followed by hormone replacement) were the tools of choice to investigate estrogen's functions *in vivo*, the cloning of the cDNA of the estrogen receptor (Walter et al., 1985; Green et al., 1986) has given first insight into protein structure and homologies compared to other steroid receptors.

Receptor binding assays are perfomed by using radiolabeled steroid-ligands, for instance tritiated estradiol. ERs are prepared from rat uterus and are solubilized and the radioligand is added to the solution. Binding occurs and the specificity is controlled by the attempt to displace the radio-ligand ("hot") bound to the receptor with a non-labeled ("cold") ligand. The receptor protein-radioligand complex is precipitated out of the solution and the radioactivitiy is counted in a scintillation counter. Through mathematical and statistical calculations the dissociation constant (Kd) for the ligand can be determined which is known for certain steroid receptors and indicates the affinity of the ligand for the receptor. The kind of graphical plotting of this binding characteristic and its displacement relationship is called "Scatchard-Plot".

A real step forward in the analysis of the estrogen receptor has been made by the crystallization of the ER, specifically its ligand binding domain (Brzozowski et al., 1997; Shiau et al., 1998). This crystallization allowed a first look into the ligand binding domain of the receptor and provided the basis for its ultrastructural description. This information is of particular importance for the search of novel more powerful antagonists and agonists of ERs and for the general specific design of drugs that target the ER. Of course this applies also to all other types of steroid receptors (for review: Fuhrmann et al., 1998).

an image of the desired mRNA a method called in-situ-hybridization is used. Here, short radio – or enzyme-labeled DNA-oligonucleotides complementary to the mRNA of the target protein are hybridized with slices of tissue, e.g. brain slices, the specific binding is made visible either by autoradiography when radiolabeling is used or by adding a substrate to the hybridized slices which is then converted by the enzyme linked to the oligonucleotide. The site of hybridization can be localized by microscopy and the slices can be semi-quantitatively evaluated. By using a combination of two oligonucleotides that are specific for two different target mRNAs and that carry different types of labels also the distribution of the mRNAs and possible co-localization with other mRNAs can be studied in one experiment (e.g. analysis of the coexpression of ERα and ERβ) (Fig. 12).
Finally, for the analysis of the expression at the protein level monoclonal or polyclonal antibodies can be used to detect the protein in protein

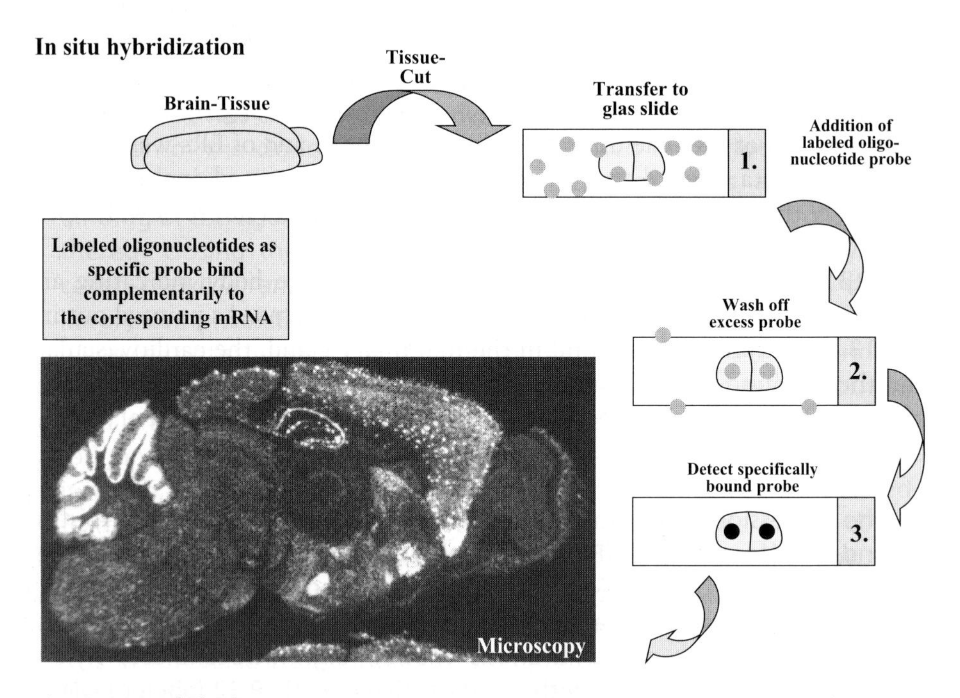

Fig. 12. In situ hybridization.
This method is used to detect the expression of the mRNA of a target gene in the tissue. Brain tissue is cut into approximately 400 µm thick slices and a DNA-oligonucleotide probe that is complementary to a stretch of the target mRNA, here the mRNA for the cannabinoid receptor 1 (CB1), is added to the brain slice. Complementary binding occurs and after the removal of excess unbound oligonucleotide probe the specific binding is detected through a label which is linked to the specific probe. Radioactive labels as well as enzymatic labels are frequently used. Here a radioactive label linked to the CB1-oligonucleotide probe depicts the mRNA transcripts for CB1 in a sagital section of an adult mouse brain (in situ-image kindly provided by Beat Lutz)

samples by Western Blotting or in the tissue (in situ) by immunohistochemistry. The first antibody specific for the desired protein is reacted with the tissue and the first antibody is recognized by a second antibody which is linked to an enzyme. The latter can convert an added substrate and a colour reaction occurs marking the site of protein expression. This method is called immunohistochemistry and it can also be used in combination with the in situ hybridizations.

There are quite some differences in the distribution of ERα and ERβ throughout the body and there is also coexpression of both receptor types in the ovary, testis, uterus, and, most interestingly, in the brain (Fig. 13). Interestingly, the mRNA of ERβ was found expressed in various regions of the brain, including lamina II-VI of the cerebral cortex and the dentate gyrus of the hippocampus (for review: Shughrue et al., 1997; 1998; Gustafsson, 1999; Shughrue and Merchenthaler, 2000).

The broad expression pattern of ERβ indeed points towards a more general role of this type of estrogen receptor in several tissues and it will be an important task to elucidate the differential physiological roles of ERβ compared with those of ERα. This knowledge is of central importance with respect to the use of estrogens as drug in estrogen replacement therapy (ERT; see page 136) or for the application of specific ER antagonists to treat estrogen-dependent breast cancers. Many expectations are lying on the evaluation of differences in the phenotype and physiology of genetically

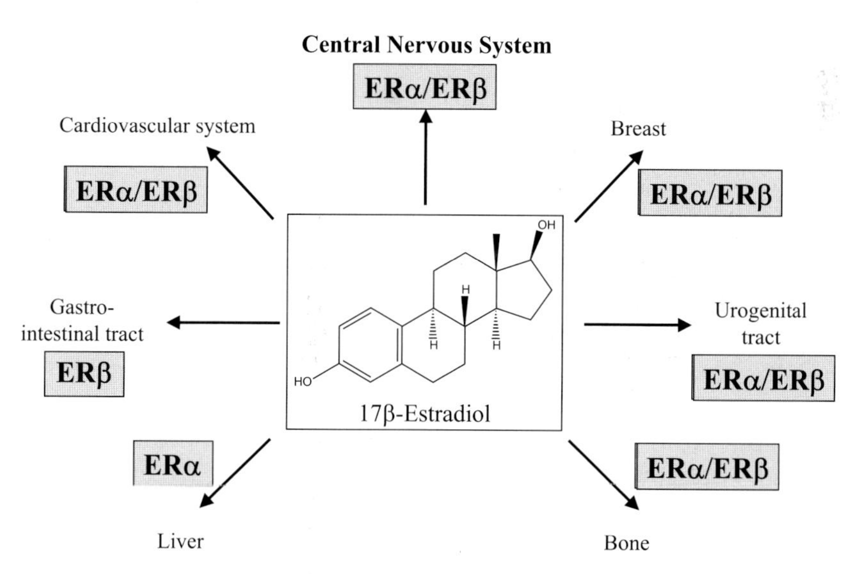

Fig. 13. Expression of ERα and ERβ in the human body.
ERα and ERβ are expressed in various tissues. In the CNS both receptors are co-expressed in certain areas of the brain (e.g. the hippocampus)

modified mice, in which the gene for ERα and/or ERβ is deleted ("knocked out"), the so-called αERKO – and βERKO – mice. The αERKO mice are available for quite some time (Lubahn et al., 1993) and much has been learned from this mouse model. In addition to the understanding of physiological functions of ERs, the clarification of the ultrastructural differences (crystal structures) of both receptor types will allow the design of specific ERα/ ERβ agonists and ERα/ERβ antagonists to differentially regulate the function of both receptors. But what are the common and what are the distinct structural characteristics of both ERs?

ERα and ERβ: a basic comparison

First of all the two types of ERs are products of two different genes. They are not two different types of one single gene, two splicing variants, a principle which is frequently occurring and which can be manifested by so-called alternative splicing. In alternative splicing, one preliminary mRNA (pre-mRNA) is processed in two ways resulting in two different mRNAs, consequently, resulting in two different proteins with possible functional differences. The molecular weight of both receptors is different and is 67 kD for ERα and approximately 54 kD for ERβ. ERα and ERβ are products of two separate genes and the gene for ERα is localized on chromosome 6 and that encoding ERβ on chromosome 14 in humans (Enmark et al., 1997).

In Western Blotting cellular proteins are separated by their molecular weight on a denaturing polyacrylamide gel. The non-visible protein pattern is transferred from the gel onto a membrane with high affinity for proteins (e.g. nitrocellulose, nylon-membrane). An antibody specific for the desired protein (e.g. monoclonal antibody from mouse specific for ERα) is reacted with the proteins fixed on the membrane and the monoclonal antibody specifically bound to the ERα is recognized by a polyclonal antibody targeted against monoclonal antibodies in general. An enzyme is linked to the second antibody and the whole binding is made visible through an colored enzyme-substrate reaction. Frequently, also enhanced chemiluminescence (ECL) and autoradiography are used to visualize the binding of the antibody to the target protein. With this method not only the presence of a protein in a protein sample can be detected but rather the molecular weight of a particular protein can be estimated. For these types of experiments frequently e.g. cells culture extracts, tissue protein preparations are used (Fig. 14).

With respect to the basic protein structure, the cloned ERβ fitted right into the steroid receptor family and shares, therefore, a common arrangement of different structural protein domains with the corresponding func-

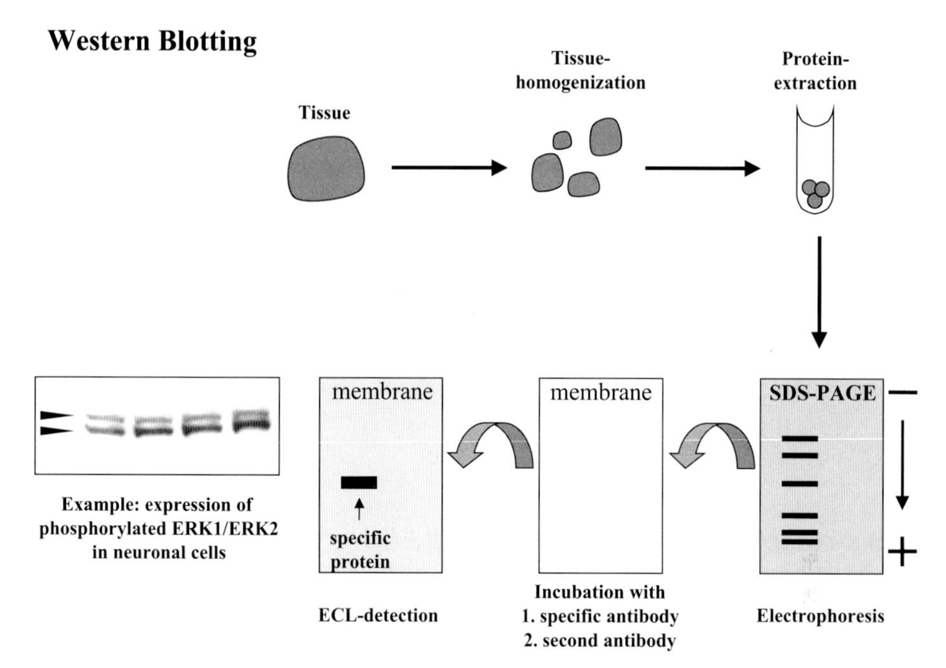

Fig. 14. Western blotting.
With this method proteins of interest can be identified in complex protein mixtures employing specific monoclonal or polyclonal antibodies. For instance the expression of a specific protein in tissues can be displayed: after the homogenization of the tissue and the biochemical extraction of proteins, the complex protein mixture is separated using a denaturing sodium-dodecyl-sulfate (SDS) polyacrylamide gelelectrophoresis (SDS-PAGE). The charged proteins migrate in a vertical electrical field and the separation of the proteins follows the molecular weight of the denatured protein. The pattern of separated proteins are transferred to filters (e.g. nylon-membrane) with a high affinity for proteins. Antibodies against the protein of interest are applied to the membrane with the immobilized proteins. The binding of the specific antibodies to the protein is visualized with a second antibody targeted against the first antibody. The second antibody is linked to an enzyme which allows the detection of the antibody binding and, therefore, the protein of interest via electrochemical luminescence

tional domains (for review: Mangelsdorf et al., 1995; Couse and Korach, 1999). The direct comparison of the protein structure of ERα and ERβ reveals that the DNA-binding domains of both show an extremely high degree of homology; in fact, the protein structure differs only in three amino acids. On the other hand there is only a 59% homology in the ligand binding domain, which is of great importance for the ligand specificity and, therefore, also of great interest for pharmacological approaches and for the search for ER-subtype specific ligands (agonists and antagonists).

The novel ERβ exists in several isoforms which are also expressed in the body indicating also a functional significance of these single isoforms rather than being molecular artefacts or experiments of the nature (Hanstein et al., 1999; Chu et al., 2000; for review: Kelley and Thackray, 1999; Gustafsson,

2000). The most prominent and best characterized of these isoforms is ERβ2 with a lower affinity for estrogen (Chu and Fuller, 1997). The existence of isoforms of ERβ with variable affinities for estrogen and possibly also for other agonists and antagonists provides the basis for a fine tuning of ERβ function and needs to be considered whenever the physiological activities of ERβ are analyzed. The exact knowledge about the isoform-specific expression pattern is also required for the understanding of the molecular interactions of ERβ with co-activators and co-repressors. For instance, it is known that ERβ 2 does not interact with the estrogen-dependent co-activator SRC-1, which very likely is the reason for the much lower affinity of ERβ 2 to estradiol (Hanstein et al., 1999). Proteins that interact with the ER and which may function as co-activators, co-repressors, anti-estrogens, and SERMs of ER function are introduced later.

The structural domains of ERα and ERβ

As with other complex proteins estrogen receptors can be divided into structural domains. These domains, by definition, describe various distinct regions in the protein molecule that may carry out certain functions or can interact with other macromolecular structures such as DNA or proteins. Of course, the nature and overall structure of these domains are determined by the amino acid sequence, the secondary, and most of all the tertiary structure, the overall folding of the protein. For instance, a groove enables another protein structure or other molecules to fit into that groove, a general interaction occurring with various ligands and the ligand binding domains of their receptors.

In case of the ERs six distinct protein domains can be defined, which are named A through F from the N-terminus to the C-terminus of the protein. The main three structural domains are the basis for the three mechanistic steps of ER action (for review: Klinge, 2000) (Fig. 15):

- The rather variable N-terminus comprises domains A and B and modulates the transcription, which is specific for a gene and a certain cell type. One specified area in that domain is the N-terminal Activation Function-1 (AF-1). Despite their structural similarities ERα and ERβ show significant differences in the N-terminal domains. According to the function of the N-terminus in both receptors it is believed that the two receptors might interact with different types of modulatory proteins.
- The domain that binds to the cellular DNA (DNA-binding domain, DBD; area C) and that comprises two so-called zinc fingers, a general structural characteristic that applies to various transcription factors.
- The so-called hinge-region or domain D separates the DNA binding domain (DBD) and the LBD and comprises a 40–50 amino acid long

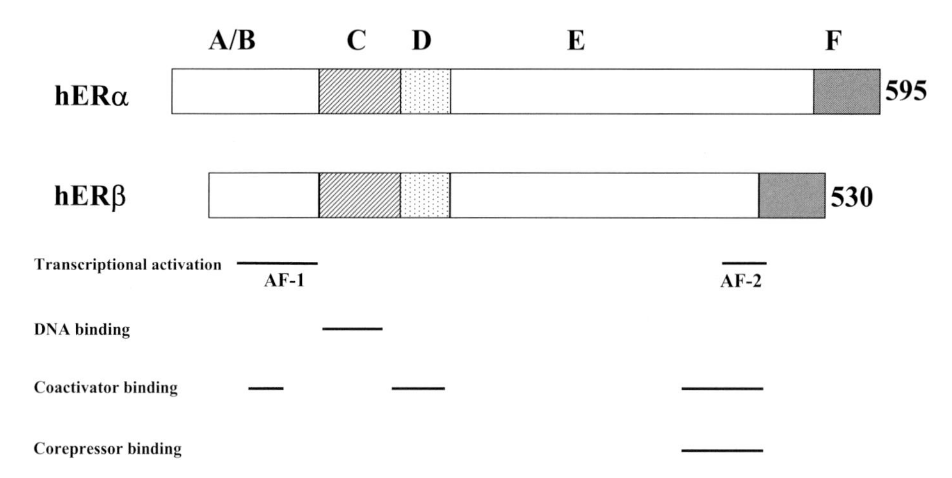

Fig. 15. Structural and functional protein domains of ERα and ERβ. (Adapted from Klinge, 2000)

peptide stretch responsible for the dimerization of the estrogen receptor and for the localization of the receptor dimer to the nucleus (NLS, nuclear localization sequences).

- The region that binds the ligand (ligand-binding domain, LBD; domain E) and which is characterized by the presence of the Activation Function-2 (AF-2). With the availability of the X-ray crystal structure of ERα a three dimensional structure of the LBD could be drawn. It could be shown that the LBD comprises 12α-helices that form a sort of pocket for the ligand. Upon binding the ligand induces a change in the protein conformation where one helix structure can form a kind of lid over this LBD pocket. This conformational change traps the ligand inside the receptor-pocket and allows other protein structures such as co-activator proteins to interact with the molecule.
- At the C-terminus of the ER the F-region is found that has been shown to play a role in the discrimination between estrogen agonists and estrogen antagonists.

By various molecular techniques including the *in vitro*-mutagenesis – a method that enables the researcher to change the nucleic acid sequences and therefore the amino acid sequence in a certain way – structure-activity relationships were further elaborated. With this technique combined with a specially designed assay system the peptide stretches or single amino acids obligatory for certain functions can be determined and protein domains can be mapped. The amino acid serine in the peptide sequence of the estrogen receptor is prone for phosphorylation and other modulations. For instance, the serine amino acid at the position 118 in the AF-1 region can be phosphorylated by the mitogen-activated-protein kinase pathway (see page 86) (Joel et al., 1998; for review: Kuiper and Brinkmann, 1994).

Phosphorylation of the ERα is a prerequisite for its activation and function. ERα is become hyperphosphorylated by its ligand after binding, e.g. estrogen. This molecular interaction triggers a chain reaction of molecular events including (1) dissociation of the ER chaperones, the heat shock proteins (mainly hsp90), (2) phosphorylation on tyrosine and serine residues, and (3) dimerization of the ER. One of the next steps is the association of the activated ER with estrogen response elements (EREs) in certain target genes of the nuclear DNA (for review: Klinge, 2000). Interestingly, the ER is the only member of the steroid hormone receptor family that is phosphorylated on both serine and tyrosine residues. Ligand-induced phosphorylation of tyrosine at a receptor are known for growth factor receptors representing tyrosine kinases, e.g. EGF-receptor. Whether this ER-typical tyrosine phosphorylation serves additional functions similar to growth factors, remains speculative so far (Kuiper and Brinkmann, 1994).

In summary: Today, two ERs that mediate the "classical" genomic mode of action are known, ERα and ERβ. Basically, both receptors are intracellular receptors for estrogen but they differ significantly in their binding affinities and ligand specificities. Both estrogen receptors belong to the superfamily of nuclear steroid receptors and show a similar basic domain protein structure. The DNA-binding domain is highly conserved (95%), the ligand-binding domain shows only a 59% homology, and there is almost no homology within the N-terminal transactivation domain. The domain structure determines the domain function of the ER (e.g. ligand binding domain, LBD; DNA binding domain, DBD). Moreover, both estrogen receptors have a different and distinct tissue distribution throughout the body.

In the following text a distinction between ERα and ERβ is made only when fundamental differences e.g. in their activities are known. Otherwise, statements concerning ERs apply to both subtypes. Prior to the identification of ERβ many genomic estrogen activities have been attributed to the ER, later called ERα, and may, indeed, rather be the result of overlapping activation of ERα and ERβ or of ERβ alone (for review: Giguere et al., 1998; Gustafsson, 1999).

Modulation of the estrogen receptor function

The majority of the studies on protein-protein interactions and the transcriptional complex performed employ *in vitro* assays, such as cell-free systems. Such studies may appear as artificial test-tube experiments when compared to the early animal experiments investigating the physiological functions of sex hormones (e.g. castration experiments). But nevertheless they show in a well-defined biochemical environment potential interactions and activities, which can occur in the living cell *in vivo*. On one hand the initial model to explain estrogen's activity via ERα as a gene switch was rather simplistic. On the other hand the actual regulation, modulation, and in particular, the fine tuning of the transcriptional activity of nuclear receptors is of immense importance and highly complex. The understanding of these regulatory events may lead to the development of compounds and drugs that block or enhance ER activity. For instance, the proliferation of certain breast cancers is dependent on the activity of ERs. ER-dependent genes in these cancer cells can be blocked by specific corepressors that have been identified. Therefore, the elucidation of steroid-protein, protein-protein, and protein-DNA interactions is of importance for the design of compounds that target ER function, such as anti-cancer drugs. In contrast to the block of ER activity, the specific enhancement of ER function via a particular coactivator could drive the transcription of potential protective genes in neuronal tissue and may lead to the development of e.g. anti-Alzheimer or anti-Parkinson drugs. Modulators of transcription factors including steroid receptors as well as other types of transcription factors are very important pharmaceutical targets for various diseases. Therefore, it can be summarized that the modulation of nuclear receptor function is not restricted to the action of agonists and antagonists but also involves the proteins active at sites outside the LBD of the receptor or at the nuclear transcriptional complex.

The crystal structure of ERα (Brzozowski et al., 1997) allowed a closer look at the physical interactions e.g. of estradiol with the ligand binding domain of the receptor. Soon the crystal structure of ERβ will also be available and a comparison of the tertiary structure of various protein domains of ERα and ERβ will increase our understanding of the structural and functional differences between these two receptors. One important first lesson of the analysis of various synthetic compounds that can modulate the activity of ERα is that it does not necessarily require a bulky molecule in order to function as ER antagonist (for review: Katzenellenbogen and Katzenellenbogen, 1996; Anstead et al., 1997). This is somewhat surprising since one may think that steric hindrance is one key to build a blockade for a receptor groove. The more that is known about functional differences and tissue specific activities of ERα and ERβ, the more effort will be put into the development of receptor subtype-specific agonists and antagonists. The *combinatorial chemistry* in sequence with *high throughput screening methods* are possible approaches for the detection of such compounds and will lead to the identification of potent pharmacological structures in the future (Meyers et al., 1999; Fink et al., 1999).

Before going into more detail about specific modulators of the ERs that interact with the receptor outside the LBD the main ligands for ERα and ERβ are summarized, and can be classified into three major categories:

- non-selective ER ligands,

including 17β-estradiol and certain ICI compounds. These ligands do not differentiate between the particular ER subtype and bind with equal affinity to both receptor subtypes,

- receptor-specific ligands,

comprising synthetic compounds e.g. (from the company Wyeth Ayerst), which specifically interacts with the ERβ subtype. High throughput screening of large samples of different designer compounds in functional ER activity assays identifies structures with the corresponding ER specificity.

- co-activator specific ligands,

such as genistein, which has a 30-fold higher affinity to ERβ compared to ERα. Genistein is an isoflavonoid derived from soy products with various activities besides its affinity to the ER. It also acts as inhibitor of protein-tyrosine kinase and of topoisomerase-II activity. Genistein is used as an antineoplastic and antitumor agent (for review: Polkowski and Mazurek, 2000; Jeffersson and Newbold, 2000).

The induction of gene transcription is a very basic and complex process which is under tight control. The requirement for the transcription of the DNA into mRNA is the assembly of various proteins including the RNA polymerase II. This assembly does not occur randomly at the DNA but is targeted to a certain region of the gene, the promoter region. There DNA stretches that are recognized by specific transcription factors, so called response elements, provide the basis for transcription factor-specific transcription. For instance, activated glucocorticoid receptors bind to promoter regions with GREs, glucocorticoid response elements, phosphorylated CREB binds to CREs, cAMP response elements, and ERs bind to EREs, estrogen response elements. Therefore, these response elements provide one basis for the specificity of gene transcription. After binding to EREs a particular assembly of various proteins occurs, which is necessary for the induction of the activity of RNA polymerase II leading to the generation of mRNA transcripts (for review: Lewin, 2000; Buratowski, 2000). There are different coactivators and corepressors that are able to join the transcriptional complex forming at the ER and many ongoing studies try to define the ultimate ER-DNA-transcriptional complex.

The EREs are the sites of interaction of the estrogen receptor protein with the DNA. This then allows the next step of ER action, the induction of gene transcription. Many genes throughout the genome carry such EREs and a minimal consensus ERE sequence has been defined, which is a 13 nucleotide-long palindromic inverted repeat (IR) with the sequence

5′-GGTCAxxxTGACC-3′.

The x-position in this sequence can be occupied by any type of nucleotide (Klein-Hitpass et al., 1988). By definition a palindrome is a DNA sequence that is identical when one strand is read left to right or the other is read right to left and, therefore, consists of adjacent inverted repeats. Variations in the affinity of the binding of the estrogen receptor to the ERE can be introduced by variations in the length of the palindromic DNA-stretch. The analysis of estrogen-regulated genes in a variety of organisms reveals that many such genes contain EREs that comprise non-palindromic and imperfect DNA-stretches (Driscoll et al., 1998). Consequently, there are quite some differences in the transcription efficiency of the different estrogen-regulated genes. One can easily imagine that a gene with a "perfect" ERE is more actively transcribed than a gene containing "imperfect" EREs or EREs with a rather scrambled sequence. Some prominent examples of human genes carrying EREs include the genes for oxytocin, previously introduced as one of the major hormones of the hypothalamus, c-fos, c-myc, known as so-called immediate early genes, and transforming growth factor-α, which is an autocrine stimulator of tumor cell growth.

In the daily laboratory work EREs are used as tools in reporter gene constructs to analyze the presence and transcriptional activity of estrogen receptors in certain cell systems. Such reporter constructs are brought into the experimental target cell by transient transfection and following the protein expression of this ERE-reporter gene construct the cells are stimulated with estrogen. If a functional ER, which recognizes and binds the reporter construct is present, the reporter gene is activated.

By recombinant DNA techniques an ERE is linked to a gene that upon activation allows its detection. Frequently used reporter genes are the genes for β-galactosidase, for CAT, and for luciferase. Using a DNA construct that combines an ERE with a reporter gene (Druege et al., 1986), here the gene for the enzyme luciferase, is very convenient since the extent of luciferase gene transcription upon activation by ligand-activated estrogen receptors can be visualized by the use of a luciferase substrate, usually luciferin. The amount of chemiluminescence generated by enzymatically modified luciferin corresponds to the level of estrogen receptor activation. The ERE-reporter DNA construct is introduced by transient transfections (Fig. 16). With this functional test system novel ER agonists and antagonists as well as the modulatory activity of coactivators and corepressors can be investigated.

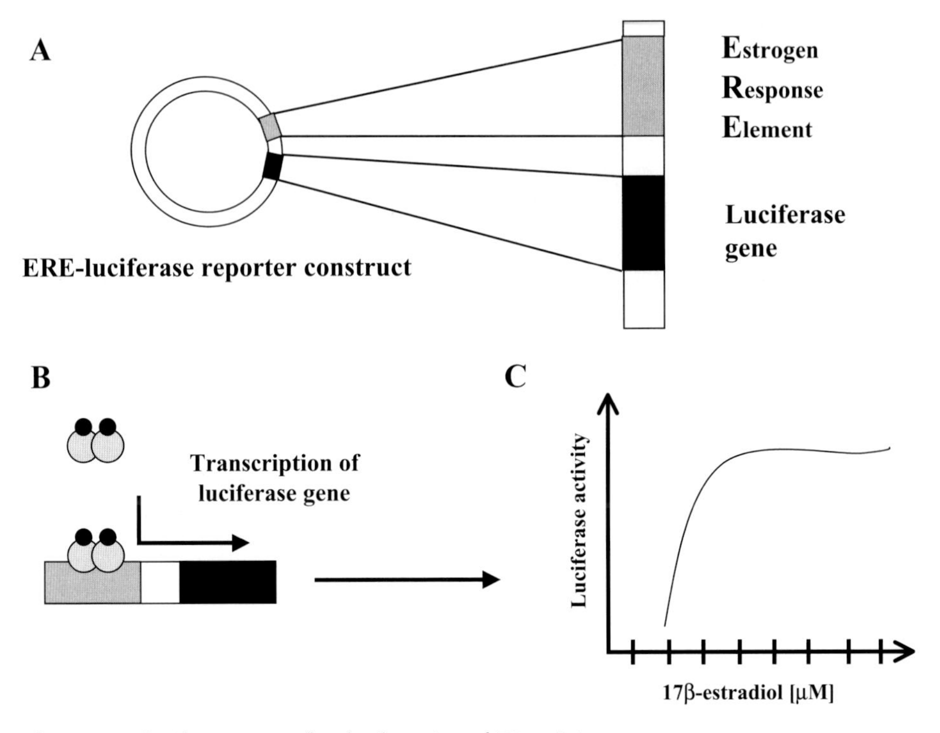

Fig. 16. ERE-luciferase assays for the detection of ER activity.
An ERE-luciferase reporter plasmid consists of ERE-DNA-stretches in the promoter region just next to the cDNA of the luciferase gene (A). Such reported constructs are transiently transferred to the cells of interest. Addition of estradiol to the cells activates the present ERs. Estradiol-bound ERs bind to the EREs on the reporter plasmid and induce the transcription of the luciferase gene (B). Extracts of the cells can be assayed for the luciferase activity which is indicative of the ER activity in these cells. A typical graph for human breast cancer MCF-7 cells treated with increasing concentration of 17β-estradiol is given (C)

This method is very powerful and allows the description of dose-response-relationships using various agonists and antagonists in cultured cells (e.g. Moosmann and Behl, 1999). Moreover, this technique can be used in a modified way and stable cell lines permanently expressing such an ERE-reporter construct can be generated. Such cell lines are used to identify novel agonists and antagonists of the classical estrogen receptor activity (e.g. Legler et al., 1999). By the modification of these DNA sequence of the EREs in the reporter construct, specific modulators of the transcriptional activation induced by an imperfect ER-DNA interaction can also be identified.

Coming back to the "classical" mechanism of gene transcription induced after binding of e.g. 17β-estradiol to estrogen receptors, translocation to the nucleus, and binding to the EREs, the next step is the assembly of the transcriptional complex and of co-activator proteins at the site of action. As occurring upon other interactions of complex molecular structures, the binding of estrogen receptors to the ERE induces conformational changes on the DNA resulting in the bending of the DNA (Nardulli et al., 1993). This physical DNA bending facilitates the next steps executed by the multiprotein transcriptional complex under the guidance of RNA polymerase II. The activity of RNA polymerase II is dependent on auxiliary transcription factors. In addition to that, before RNA polymerase II gets into action a rather complex mixture of proteins of the so-called TF-family (e.g. TFIIA, TFIIB etc.) needs to be assembled at the core promoter site. TF-proteins are general factors. This assembly establishes the basal transcription apparatus. All these proteins interact in a certain order leading to conformational changes. As mentioned before this represents a basic concept in biology, structure e.g. protein conformation determines biological function. Steroid receptors including estrogen receptors interact with the basal transcription factors TFIIB, TBP, and others (for review: Klein-Hitpass et al., 1998). These additional proteins are household proteins and are present in every cell. After assembly of the complete transcription initiation complex that is triggered by binding of estrogen receptor binding to EREs, RNA polymerase II is entering the transcription start site (ATG) and launches the transcription of the ER-controlled mRNA. For a complete overview over the transcription process the reader is referred to the textbook (for review: Lewin, 2000).

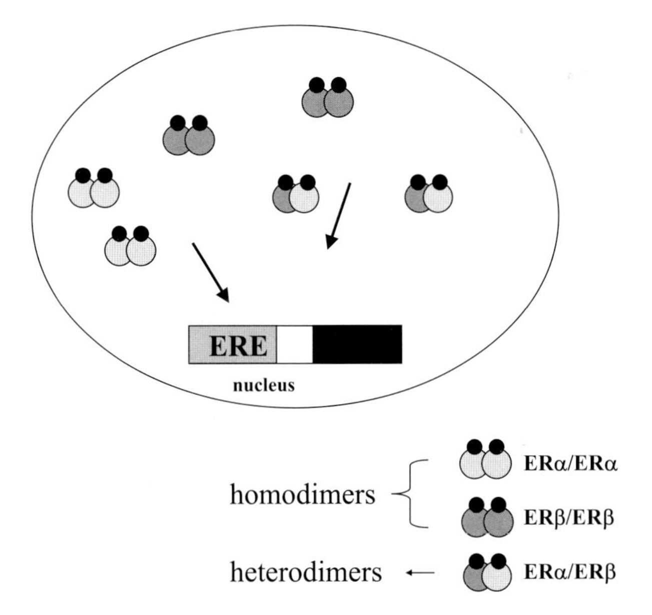

Fig. 17. Dimerization of ERα and ERβ.
ER dimers interact with EREs. Homodimers and heterodimers can be formed, which may target different genes

Interaction of the ER with co-activators and co-repressors: of RIPs, RAPs and DRIPs

Like many nuclear receptors the ER (here most of the data have been acquired for ERα) also directly interacts with co-activators and enhance transcription. It is far beyond the discussion in this book to review in detail the many different co-activators identified so far. But it should be mentioned here that co-activator proteins are also called receptor interacting proteins (RIPs) or receptor associated proteins (RAPs), which makes clear that these proteins interfere already at the level of the receptor and not at the DNA. ERα interacts with different classes of such co-activating proteins (for review: Horwitz et al., 1996; Klinge, 2000). Recent studies uncovered yet another family of proteins which are able to interact with ERα, the so called DRIPs (vitamin D receptor interacting proteins). DRIP is a novel multi-subunit coactivator complex required for transcriptional activation by nuclear receptors and other transcription factors. Interestingly, the binding of the protein DRIP205 (205 corresponds to the molecular weight of 205 kD) to ERα and ERβ is dependent on the activation of the receptor by its ligand 17β-estradiol (Burakov et al., 2000; for review: Freedman, 1999; Rachez and Freedman, 2000). While certain ER agonists induced the ER-DRIP205 interaction, this interplay is inhibited by antagonists of ERs such as tamoxifen and raloxifen. Other ER interacting proteins and co-activators are SRC1, SRC3, and CBP3 (for review: Klinge, 2000). Therefore, the "classical" model of estrogen and ER action at the DNA (see Fig. 11) needs to be revised. Up to now more than 30 different adaptor proteins have been described. One prominent candidate, the steroid receptor co-activator-1 (SRC-1), was already previously mentioned and is intensively studied (for review: Leo and Chen, 2000). A model of ER action which now includes the concept of adaptor proteins is given in Fig. 18. The central importance of co-activators is nicely demonstrated in a recent animal study demonstrating that SRC-1 protein expression is critically involved in the hormone-dependent development of normal male reproductive behavior and brain morphology (Auger et al., 2000).

The molecular counterparts of the co-activators are the co-repressors. Concerning the latter the available information is much more rare. Indeed, only a few reports are available on the interaction of ERs with co-repressors. Co-repressors negatively influence ER activity and may, ultimately, also be important targets for pharmaceutical intervention. Of course, co-activators and co-repressors play a major role in the modulation of the ER target gene expression in a tissue-specific manner. The positive and negative fine tuning is mediated by the expression of different levels of these co-activator and co-repressor proteins depending on the particular cell type as shown for the expression of the co-activators SRC-1, RIP-140, p300, and CBP and for the

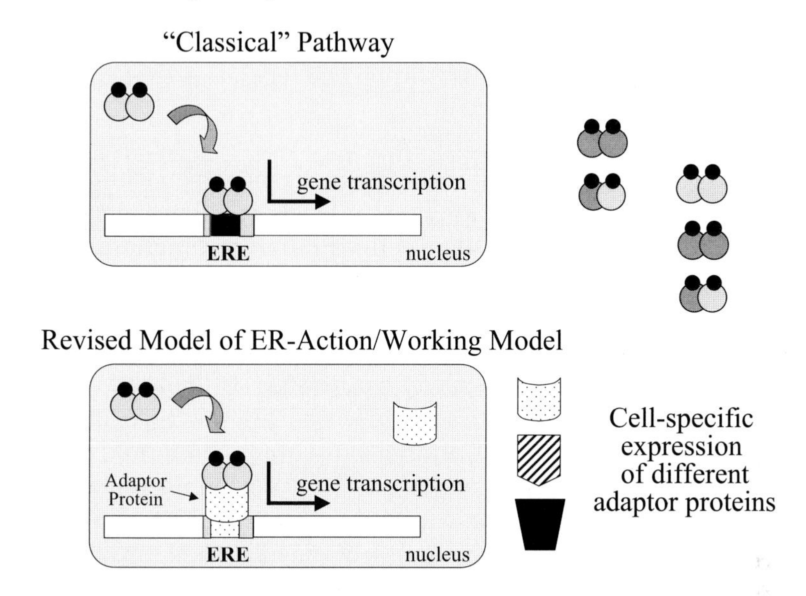

Fig. 18. Revised model of ER-action: working model.
The events occurring in the nucleus of ER-expressing cells are depicted in the "classical" view and in form of a revised model. The revised model considers adaptor proteins that mediate the binding of the ER-dimers to the EREs. The cell specific expression of certain adaptor proteins may cause the tissue-selective expression of ER-regulated genes

expression of co-repressors including SMRT and NCoR in rat tissues and various cell lines (Misiti et al., 1998; Folkers et al., 1998; for review: McKenna et al., 1999). To put it very simply: if lots of co-activator protein is present in a cell a high ER-driven transcriptional activity can be expected and vice versa (for review: Collingwood et al., 1999; Klinge, 2000; Leo and Chen, 2000).

Before the discovery of the second estrogen receptor, ERβ, one main focus of estrogen research was the elucidation of estrogen target genes in various tissues and the molecular analysis of estrogen-responsive promoters. With the identification of ERβ the tissue specific expression pattern of both receptors and the identification of ER subtype-specific co-activators and co-repressors is in the center of interest. Therefore, the question of how both receptors are differentially modulated needs to be answered. In addition, the analysis of possible coexpression patterns of ERα and ERβ in certain tissues that would also allow the heterodimerization of both receptors is of great physiological and pathophysiological significance. If one considers also the putative existence of a third or fourth ER subtype and also the confirmed presence of various orphan receptors that, although lacking a ligand, may interact (e.g. dimerize) with the known ERs, it becomes evident that the number of possible modes of action is increasing. Homodimers of ERα and ERβ may induce the transcription of other estrogen target genes than heterodimers of ERα and ERβ. As confusing as this may seem at the first site common features of ER activation and regulation that apply to both ER subtypes are well known.

In summary, the genomic activity of ERs as ligand-dependent transcription factors can be modulated at different levels as a consequence of the specialized domain structure that defines different functional sites at the ER. By putting all above data into context it appears that the activity of the ER transcription factor (and this applies for both types of ERs) can be modulated and controlled at various levels. There are several checkpoints for the control of the activity comprising (1) the level of expression of ERs in various tissues themselves, (2) the type of agonist (17β-estradiol, estrol, estrone) or antagonist (tamoxifen, raloxifen, ICI antagonists), (3) the particular way of dimerization (homo-versus heterodimerization), (4) the specific structural feature (length of the DNA stretch and composition of nucleotides) of the EREs, and, ultimately, (5) the ratio of expression of co-activators and co-repressors in one particular cell. Of course modulators of ER action, in general, and co-activators and co-repressors, in particular, are important pharmaceutical targets for drugs that aim to modulate ER activity in a tissue-specific manner. The tissue selectivity would avoid some side effects of drugs. For instance, blockade of the ER, which drives the growth of breast cancer cells, by a non-selective ER antagonist would also block ERs in the brain, where another antagonist may serve basic neuromodulatory function. Therefore, the key issue of using ERs as therapeutic target is the tissue-selective and -specific modulation of ER function. The type of ER activity described here, which is mediated via the induction of gene transcription caused by ERs, is called the genomic or the "classical" mode of estrogen and estrogen receptor activity.

For a long time this "classical" mode of estrogen receptor action has been believed to be the only one possible. But as already pointed out shortly above and as it will be described in more detail in the following paragraphs, activated ERs can also interact with signaling pathways and, therefore, indirectly modulate gene transcription without binding itself to an ERE-containing target gene. The question is still open whether there are more types of estrogen receptors besides ERα and ERβ. Various experimental observations could not be interpreted with the "classical" ER activities and other types of ERs, membrane ERs, have been postulated but have not yet been cloned. Some experimental evidence for the existence of a membrane ER will be introduced below. Next, the topic of the selective ER modulation is complemented by introducing the term SERMs, selective estrogen receptor modulators.

Selective estrogen receptor modulators – SERMs

As discussed the term SERMs comprises a variety of proteins which (1) are found to be associated with the ER in the nucleus and (2) can function as co-activators or co-repressors. The range of activities of SERMs rapidly increased the level of complexity of ER function. Selective ER modulators (SERMs) may answer the important question of tissue-selective activities of estrogen and other ER ligands. A tissue-specific activity was observed for the drug tamoxifen, which, historically, is the first described SERM. Today's knowledge about the biochemical basis of ER function suggests that upon binding of a ligand to the ER, a conformational change in the protein structure of the ER is occurring. This conformational change may have functional consequences since it determines the accessibility of the receptor protein for certain adaptor proteins to associate in a specific manner (e.g. Levenson and Jordan, 1998). This particular association then in turn may affect the downstream activities of the ER at the DNA in the nucleus. It is therefore of central importance to determine the expression levels of such central adaptors in an estrogen target cell expressing ERs (for review: Osborne et al., 2000; McDonnell, 2000).

Among other tissues throughout the body estrogen acts in the cardiovascular system, skeletal bone, and the nervous system. Since ERs are widely expressed, one pharmacological goal is to develop ER ligands with a high tissue selectivity. A specifically designed highly selective ER ligand should only exert its agonistic or antagonistic function in the desired target tissue and not in others, where it would have untoward effects. Although many different beneficial activities of estrogen intake as in estrogen replacement therapy (ERT) in postmenopausal women have been reported, the compliance and the acceptance of ERT is still only very limited, mainly because of unwanted estrogen effects. The strongest anti-ERT argument is the fact that in many studies an increased likelihood of developing breast cancer is reported although as many studies did not find such an effect. This issue is still under discussion (Beral et al., 1997; Lacroix and Burke, 1997; see page 138). SERMs could be a solution for this dilemma. Tissue-selectivity has to be the key feature of SERMs. Again, SERMs could be developed with the ability to modulate ERα activity towards the transcription of neuroprotective genes in the brain and no gene activating properties within the bone or breast. Such SERMs would be of great clinical significance. Tamoxifen is a prominent example of a SERM although initially thought to be a pure ER antagonist (for review: Jordan, 1999). Tamoxifen has been used clinically for more than 20 years to treat breast cancer. The strategy here is to antagonize the ER in breast tumor cells whose growth is dependent on estrogen and ER activity. ERs were introduced in the previous chapter as a trigger that switches on transcription in a very simplified model (see Fig. 11). Anti-estrogens such as tamoxifen can block this switch and, therefore, can prevent estrogen function. Interestingly, step by step it became apparent that tamoxifen is rather a SERM instead of a pure antagonist. It was found that tamoxifen exerts ER-antagonistic effects in breast cancer cells on one hand and on the other has ER-agonistic estrogenic function in skeletal and cardiovascular tissue. Love and coworkers initially showed that tamoxifen did not act as an anti-estrogen in the bone but rather is a partial agonist (Love et al., 1991, 1992, 1994; Sato et al., 1996; for review: Love, 1995).

After these initial findings with tamoxifen other SERMs have been identified including raloxifen, which is also in clinical use. One further complication with respect to SERMs arose after the cloning of ERβ in 1996 since the ligand binding domain of ERα and ERβ shows quite some differences in the sequence and structure. Consequently, cross-reactivity of certain SERMs with the ER subtypes needs to be determined. Indeed, estrogen signaling and physiology and ER pharmacology needs to be re-evaluated after the discovery of ERβ. Although estrogen (17β-estradiol) may bind with similar affinity to both receptors, differences in the affinities of SERMs are found, a fact which has to be taken into account for a rational design of SERMs.

But still, the identification of a second ER does not provide a sufficient explanation to explain the pharmacological properties of SERMs, such as that different members of the SERM family of compounds can have quite different physiological effects in one and the same cell. The solution of this problem was delivered by different studies analyzing the fine structure and biochemistry of the ER-SERM complex. There it was found that there is quite some difference in the conformation of this protein complex when occupied by tamoxifen or raloxifen, other SERMs, or 17β-estradiol (Shiau et al., 1998; McDonell et al., 1995; Brzozowski et al., 1997; Pike et al., 1999; for review: Anstead et al., 1997). Obviously, the specific type of conformation of the ER-ligand complex determines whether there is an agonistic or antagonistic activity. And indeed, it was found that at the DNA there is no direct physical interaction of the ER with the DNA as previously believed but rather that this interaction is mediated via specific adaptor proteins, whose expression is different in various tissues and cells (for review: Horwitz et al., 1996) (see Fig. 18).

A variety of compounds that bind to the ER display an activity quite different from the "classical" estrogens such as 17β-estradiol. Based on the knowledge that the growth of certain types of breast cancer is ER-dependent for a long time the most important feature of an ER modulator was to antagonize and block ER action. With the identification of the wide range of ER expression and function selective ER modulation is the aim. It has been recognized that some of the initial anti-estrogens are indeed selective estrogen receptor modulators (SERMs). SERMs bind to ERs, change their conformation, and depend on this particular change. Adaptor proteins, such as SRC-1, can associate with the receptor before activation of ER-responsive target genes. Adaptor proteins may act as co-activators or as co-repressors of ER activity and the expression of different adaptors can be different in various tissues. This may be one key to the tissue-selective differential activity of SERMs.

To be or not to be? Are there membrane ERs?

As discussed in the previous chapter, estrogen's activity via ERs is regulated at different levels involving two types of ERs and various co-factors, SERMs, and adaptor proteins that may mediate the fine tuning and the tissue-specificity of ER action. At this point, the mechanisms of ER activities appear rather complicated and much more information is necessary to understand the full picture. The mode of action of estrogen via the activation of ERs, the "classical" mechanism, induces a genomic response with long-term consequences. Since gene transcription takes some time this effect of estrogen is frequently also called "delayed effect". In contrast to this delayed signaling, estrogen may also induce very rapid cellular responses, which are then consistently called "rapid effects".

Rapid and "non-classical" effects of estrogen involve direct molecular interactions of the steroid with the cellular membrane influencing its fluidity or with integral membrane proteins, which frequently represent ion-channels. These non-genomic and rapid effects will be in the center of the next chapter. But before that, a putative form of ER, which is believed to also be able to induce rapid intracellular effects, is discussed here shortly, the possible membrane ERs. The idea of the existence of membrane ERs is very intriguing but this issue is still under debate. Do membrane steroid receptors such as membrane ERs really exist? And if they exist, what is their structure?

To begin with, membrane receptors for estrogen have been postulated at least since 1977 (Pietras and Szego, 1977; Pietras and Szego, 1980) and, therefore, rather long before the cloning of the "classical" ERs. And there is no doubt that not only 17β-estradiol but steroids in general can directly interact with the cellular membrane and also with protein structure (receptors, ion-channels) that are embedded in or associated with the cellular membrane. But some of these observations, especially the early findings (Pietras and Szego, 1977) from today's standpoint may be explained by the activity of estrogen as a *neuroactive steroid* (see page 79) rather by binding to or activation of a specific receptor. The biggest drawback of the hypothesis of the existence of membrane ERs is the fact that little is known concerning their putative biochemical and molecular structure. Initially, there has been some success in a partial purification from membrane ERs from hepatocyte membranes (Pietras and Szego, 1980), nice work, which was then, unfortunately, never really followed up on. Nevertheless, it is believed that the membrane and the intracellular ER may have a quite similar amino acid sequence and epitope structure since antibodies specific for ERα also recognize proteins attached to or integrated into the cellular membrane (Pappas et al., 1995; Morey et al., 1997).

In recent observations from the laboratory of Ellis Levin it was shown that using cultured chinese hamster ovary (CHO) cells the transfected ERα cDNA is translated into an intracellular ERα and, very likely following post-translation modification, also into a membrane ERα. To demonstrate this bovine serum albumin chemically linked to estradiol and to a radioactive label was used and found attached to the cellular membrane. In fact, in this set of experiments, also the binding of labeled progesterone to a specific site at the cellular membrane was observed. Therefore, the observed membrane activity was not restricted to estradiol (Razandi et al., 1999). Consequently, it may be asked whether there are also membrane progesterone receptors? However, the major support for the existence of a membrane ER comes from studies that present direct evidence for rapid, non-genomic, non-classical activities of E2. The modulation of intracellular signaling events can be induced by E2 in seconds to minutes and involves various intracellular messenger systems including the release of Ca^{2+} from intracellular stores, activation of adenylate cyclase or phospholipase C and also the extracelluar-regulated-kinase (ERK)/mitogen-activated-protein-kinase (MAPK) pathway (Aronica et al., 1994; Tesarik and Mendoza, 1995; Le Mellay et al., 1997; Migliaccio et al., 1996; Watters et al., 1997). Indeed, the interactions of E2 with the intracellular signaling are of major importance since they modulate the translation of extracellular signals into intracellular information. Later in the book the effects of estrogen as neuroactive steroid and its direct "cross talk" with intracellular signaling events will be discussed with respect to neuroprotection. Interestingly, membrane binding sites for estradiol have also been demonstrated in osteoblast- and osteoclast-like cells and in cultured neostriatal neurons and in the rat brain (Fiorelli et al., 1996; Mermelstein et al., 1996; Ramirez and Zheng, 1996). For a more detailed review on evidence pro membrane ERs the interested reader is also referred to a recent review (for review: Levin, 1999).

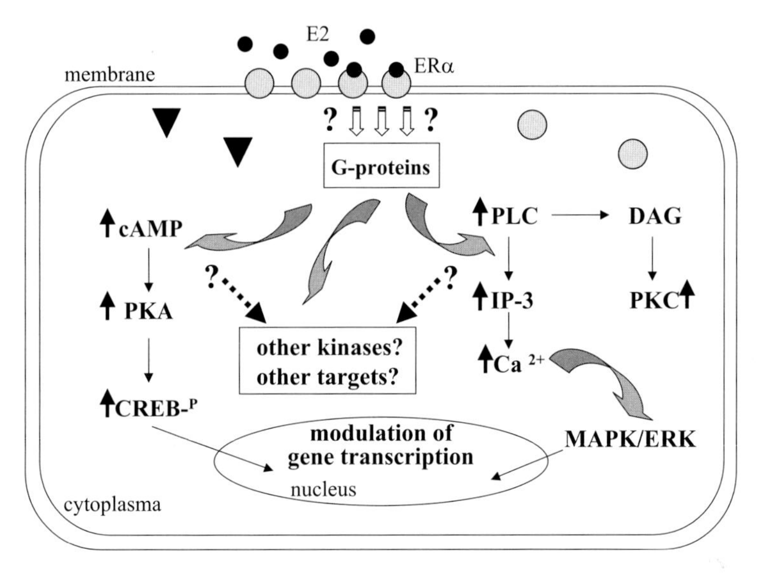

Fig. 19. Estrogen membrane receptors activate intracellular signaling.
Membrane ERs may interact with various intracellular signaling pathways. Similar to classical protein hormone receptors membrane ERs transduce the estrogen signal into the interior of the cell via the activation of second messenger molecules. Central pathways including the cAMP/CREB-pathway and the MAP Kinase pathway may, ultimately, cause the modulation of gene transcription (modified after Levin, 1999)

In conclusion, the evidence for the presence of a membrane ER is more or less of indirect nature and only the purification and cloning of this possible receptors (if its a new one and not only a modification of the known ERα and ERβ) will convincingly show their presence and will allow further studies.

4. "Non-classical" activities of estrogen

It has been repeatedly mentioned that steroid hormones modulate gene transcription through cognate intracellular receptors, which is defined as the "classical" mechanism. The induction or repression of gene expression cause a long-term genetic and, ultimately, physiological response. The response is not immediate but delayed. Experimentally, this complex process of steroid-induced gene transcription can be blocked at different stages. To do so, mostly steroid receptor antagonists are applied that prevent the binding of the natural ligand. Frequently, also compounds are used that inhibit processes more downstream such as transcription of the ER-specific mRNA or translation of the mRNA into protein. Actinomycin D and cycloheximide are frequently used in cell culture experiments to inhibit the latter processes. These series of events are called "classical", genomic, or delayed since the response takes some time due to the protein syntheis occurring in the cell (transcription and translation).

Rapid non-genomic effects compared to slow genomic effects of steroid hormones: what makes the difference?

In addition to the delayed genomic response, in many experimental paradigms steroids have been shown to exert also very rapid activities. As already heard it appears that all types of steroids may have also a rapid, non-genomic mode of action and that, therefore, this ability is not restricted to the female sex hormone. The requirement for such an activity may be based in the chemical structure; steroids are lipophilic compounds.

But what means "rapid" with respect to cellular physiology? Some genomic effects may occur also rather rapidly. Among the most rapid genomic steroid effects is the glucocorticoid-induced transcription of certain viral sequences (long-term-repeats), which may occur as early as after 8 minutes (Groner et al., 1983). But usually the steroid receptor-mediated induction of transcription is slower and takes up to several hours. This delay is much shorter during the non-genomic steroid effects. Estrogen, for instance, modulates intracellular signaling inside the cells and vasoregulation within 1 to 2 minutes (Christ et al., 1995; Hammerschmidt et al.,

1988; Gilligan et al., 1994; Manegold et al., 1999). Further, these rapid onset effects even occur despite the inhibition of transcription and translation by actinomycin D and cycloheximide respectively or despite the presence of steroid receptor agonists suggesting a non-genomic steroid effect. Such rapid effects of steroids have been investigated for quite some time and there was the early observation that progesterone has an immediate anaesthetic effect in certain experimental models (Seyle, 1942).

Rapid effects of estrogen

The female sex hormone estrogen is also very intensively investigated with respect to rapid steroid effects. In 1975 it was found that estrogen induces the influx of calcium into myometrial cells, cells of the uterine smooth muscles (Pietras and Szego, 1975). Similar rapid estrogen responses were seen also in other tissue and cells including pituitary cells, human oocytes, granulosa cells, and neostriatal neurons. Other intracellular target systems for the rapid non-genomic estrogen effects are also described. For instance, in vascular smooth muscle cells (VSMC), breast cancer cells and myocytes estrogen rapidly alters the intracellular level of cAMP (Li et al., 2000; for review: Schmidt et al., 2000). Interestingly, the concentration range in which these rapid activities of estrogen are reached in the physiological range (nanomolar concentration). Many findings on estrogen's rapid effects are derived from studies focusing on estrogen's well-known cardioprotective function (for review: Mendelsohn and Karas, 1999). 17β-estradiol and its derivative ethinyl-estradiol have significant and clinically relevant relaxing effects in coronary artery tissue from pig and humans as well as in rat aortas (Salas et al., 1994; Rodriguez et al., 1996; for review: Schmidt et al., 2000). This estrogen effect is of great clinical importance for atherosclerosis, a disease of the cardiovascular system which is characterized by the narrowing and hardening of the arteries. For a number of these activities the existence of membrane estrogen receptors has been claimed and it is hypothesized that such receptors may be directly linked to intracellular signal transducers, including G-proteins or adenylate cyclase (Fig. 19).

However, evidence is now collected demonstrating that the intracellular classical receptors (ERα and ERβ) may also take part in the rapid steroid actions. Estrogen-induced calcium-dependent stimulation of nitric oxide synthase (NOS) release in cultured endothelial cells from ovine pulmonary artery, which occurs within minutes, could be blocked by ER antagonists such as tamoxifen and ICI-182,780 (Lantin-Hermosos et al., 1997; Shaul, 1999). The enzyme NOS catalyzes the conversion of L-arginine and oxygen to nitric oxide (NO), which is the physiological vasorelaxant. Interestingly, this estrogen effect was not prevented by actinomycin D (Chen et al., 1999), indicating that no mRNA transcription is involved in this process. These data further suggest an additional mode of action of ERs. The ER-ligand complex may interact directly with intracellular enzymes (e.g NOS) and may modulate the enzymatic function. It may be speculated that ERs are associated or complexed with certain intracellular enzymes including NOS. Upon activation of the ER with estrogen, which is a rapid process by itself, the protein structure of the entire protein complex changes leading to the modulation of NOS activity. Therefore, membrane receptors would not necessarily be needed for estrogen to reach a rapid activity. Physical interactions in form of protein complexes have been recently shown for ER and phosphatidyl-inositol-3 (PI-3) kinase (Simoncini et al., 2000). Moreover, caveolar structures, which are regular invaginations of the cellular membrane, may play a role (see page 89; Fig. 31).

In addition to estrogen's rapid effects on cAMP-synthesis, estrogen very rapidly induces tyrosine phosphorylation and activation of the mitogen-activated-protein kinases (MAPK) ERK-1 and ERK-2, an effect that can not be blocked by ER antagonist (Manthey et al., 2000). MAPK-signaling is believed to also play a major role during cellular neuroprotection. These data and a discussion of the consequences of this interaction for nerve cell protection and for neurodegenerative disorders is presented later. For a complete review of the literature on rapid non-genomic effects of steroids in general, the reader is referred to previous publications (for review: Paul and Purdy, 1992; Baulieu, 1997; Revelli et al., 1998; Schmidt et al., 2000).

Structure-dependent effects: estradiol as antioxidant

Reactive oxygen species (ROS): normal byproducts of life under oxygen

There is yet another role that estrogen can play: estrogen can also act as an antioxidant. What is an antioxidant good for? Antioxidative molecules scavenge free radicals. And free radicals are molecules with unpaired electrons conferring a high and detrimental reactivity to this molecule. Antioxidants counteract the destructive effects of oxidations in tissue and cells. Free radicals are obviously very unstable molecules and in order to acquire a more stable energy level, free radicals can accept electrons or a hydrogen atom from another molecule. With the oxidative phosphorylation reactions that build the center of the respiratory pathway in the mitochondria of cells, molecular oxygen is reduced to water (H_2O) by the successive acceptance of electrons and protons. During, this process three main reactive oxygen species (ROS) are formed: superoxide ($O_2{\cdot}^-$), hydroxyl radicals ($OH{\cdot}$), and the intermediate hydrogen peroxide (H_2O_2). Of course, H_2O_2 is chemically not really a radical molecule but it is the substrate for the formation of highly reactive hydroxyl radicals. This conversion occurs in the presence of Fe^{2+} via

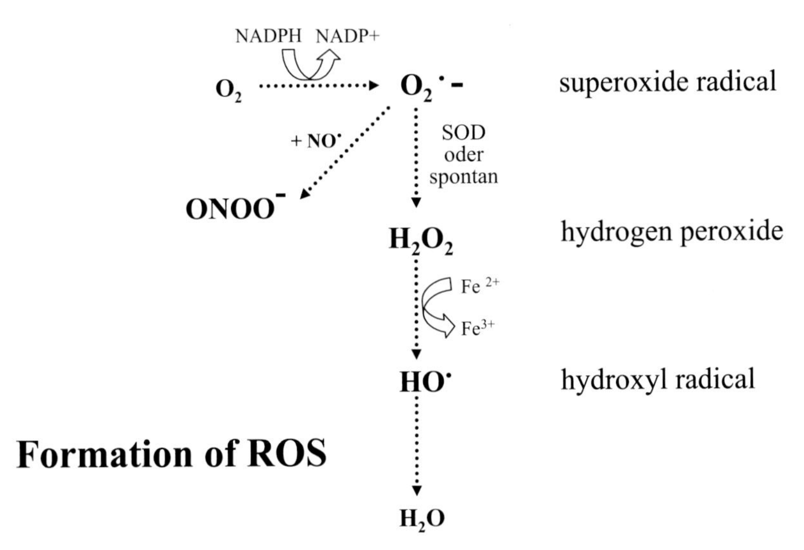

Fig. 20. Formation of reactive oxygen species.
The main reactive oxygen species (ROS) in mammalian cells are superoxide radicals, hydrogen peroxide, and the hydroxyl radical. Electrons from the electron transport chain in the mitochondria of the cell may falsely find their way to an oxygen molecule, which then is dismutated to H_2O_2 via the superoxide dismutase (SOD). In the presence of catalytic ions, e.g. Fe^{2+}, the highly aggressive hydroxyl radical is generated. The latter attacks electrons present in the carbohydrate side chains of the lipids of the cellular membrane causing lipid peroxidation (see Fig. 23). While regularly ROS are produced in the cells which are then detoxified by antioxidant enzymes, the accumulation of ROS may cause oxidative damage to the biomolecules of the cell. In nervous tissue various mitochondria-independent enzymatic and non-enzymatic mechanisms may cause the formation of ROS (see Fig. 33)

the so called Fenton-Reaction. Iron is an important catalyst of oxidation reactions and also promotes the generation of ROS (Aust et al., 1985; Sies, 1993; Hippeli and Elstner, 1997; Halliwell and Gutteridge, 1999).

When biomolecules are modified by oxidation, frequently loss of function or dysfunction is the consequence. The series of pathological events induced by oxidation which can range from only minor changes in gene transcription up to complete cell destruction, is called oxidative stress. Due to the broad range of molecules that can be oxidatively modified (DNA, proteins, lipids) oxidative stress is strongly believed to play a major role during a variety of pathological processes ranging from human cancer to the formation of cataracts, skin damage, arteriosclerosis to neurodegenerative disorders such as Parkinson's Disease and Alzheimer's Disease (for review: Sies, 1993; Beckman and Ames, 1998; Halliwell, 2000). Of course, we are permanently under some kind of oxidative stress. ROS are natural byproducts of our life which during the evolution developed in the oxygen atmosphere. There is a general equilibrium of the production of ROS and the detoxification of ROS which is mediated by various antioxidant defense systems. Indeed, the sources of ROS generation are numerous in the tissue and cells including nerve cells. Especially those body tissues which do contain cells with a rather limited or no regenerative potential, such as neuronal cells, have to maintain an effective anti-oxidation defense system in order to protect themselves against detrimental free radical damage (for review: Sies, 1993).

Antioxidant defense lines of the cell

Enzymatic and non-enzymatic antioxidants keep the fine-tuned balance between the physiological production of ROS and their detoxification (for review: Sies, 1997). The main physiological source of ROS is the sequence of reactions during oxidative phosphorylation, necessary for the generation of adenosine triphosphate (ATP), the central energy source of nerve cells in the brain. Oxidative phosphorylation occurs within the defined micro-environment of the mitochondria and in this compartment the free radicals are quickly reduced to H_2O. Since nothing is really perfect in this world and this applies also to physiology, – but in fact errors drive the evolution – leaking

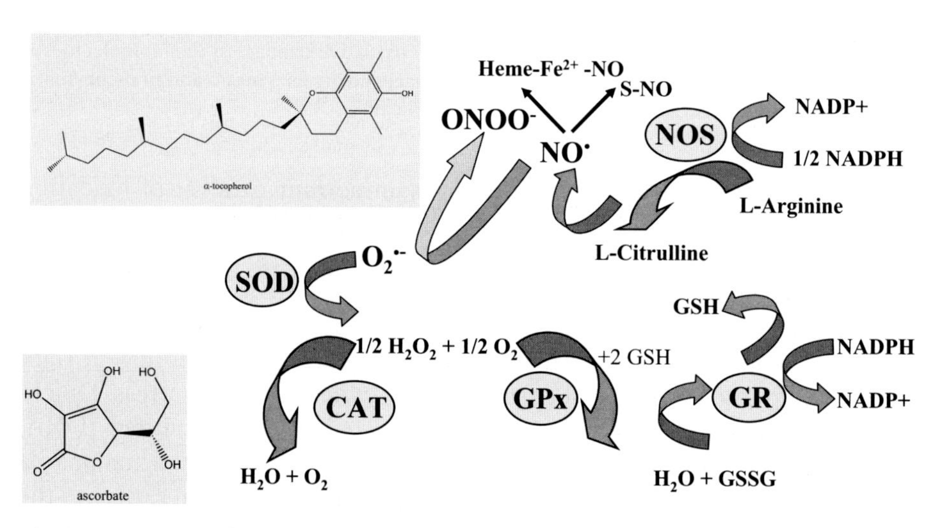

Fig. 21. Enzymatic and non-enzymatic antioxidant defense.
Various antioxidant enzymes detoxify accumulating ROS. Superoxide dismutase (SOD) converts superoxide radicals and generates H_2O_2. Catalase (CAT) and glutathione peroxidase (GPx) convert H_2O_2. GPx acts via the reduction of the tripeptide glutathione (GSH). Glutathione reductase (GR) then recycles the GSH. Nitric oxide (NO) is produced from the conversion of L-arginine to L-citrulline catalyzed by nitric oxide synthase (NOS). Peroxynitrite ($ONOO^-$) can be formed from superoxide radicals and NO. The most prominent non-enzymatic antioxidants are α-tocopherol (lipophil) and ascorbate (hydrophil)

$O_2^{\cdot-}$ can be dismutated by the enzyme superoxide dismutase (SOD) to H_2O_2. Then, the latter is a substrate for the intracellular antioxidant glutathione (GSH) and for the enzymes catalase and glutathione peroxidase (GSH-Px). SOD, catalase, and GSH-Px are enzymatic acting antioxidants. Other antioxidant molecules are of non-enzymatic nature and are non-enzymatic and chain-breaking antioxidants. The lipophilic free radical scavenger α-tocopherol (vitamin E) and the hydrophilic ascorbate (vitamin C), are the two most prominent non-enzymatic antioxidants. They directly react with ROS at the molecular level. In addition to its direct reaction with ROS, ascorbate is also necessary to regenerate vitamin E (for review: Sies, 1993; Sies und Stahl, 1995) which is a chemical reaction of great physiological importance (Fig. 21).

An increased production of free radicals, which means a misbalance of the oxidant-antioxidant-system, either induced by an overdrive of endogenous ROS generating systems or by exogenous oxidative insults leads to oxidative stress with its fatal consequences (Olanow, 1993; Ames et al., 1993; Coyle and Puttfarcken, 1993; Sies, 1997; Behl, 1999). It has to be added here that ROS also have central physiological functions including the modulation of intracellular signaling via redox-sensitive transcription factors, e.g. nuclear factor kB and activator protein 1, AP-1, a heterodimer between the proteins Jun and Fos, and the destruction of pathogens as intruders into the body (for review: Sies, 1997; Halliwell and Gutteridge, 1999). The ability of cells to respond to changes in the oxygen milieu is vital for cells. The redox-sensitive transcription factors are in charge of counteracting upcoming potential dangers via the induction or repression of gene transcription.

Estradiol is an antioxidant similar to α-tocopherol (vitamin E)

Estrogen is a steroidal compound and chemically 17β-estradiol, the bioactive estrogen, belongs to the group of aromatic alcohols due to its phenol group in ring A. The proton in the hydroxyl group of the aromatic ring A can be donated to free radicals that carry an extra non-paired electron. This leads to the detoxification of the free radical and to the generation of an antioxidant-radical (Fig. 21), which then can be recycled by various reactions (for review: Halliwell and Gutteridge, 1999). This is the basic mechanism of free radical scavenging by phenols and holds true not only for vitamin E (α-tocopherol), which is a well-known lipophilic antioxidant with a phenolic structure, but also for 17β-estradiol. Indeed, the basic structure of vitamin E and estradiol is comparable since both compounds consists of a hydroxyl group at a mesomeric ring system (antioxidant moiety) and a carbohydrate tail (lipophilic moiety).

Employing various cell-free *in vitro* paradigms of peroxidation reactions, it has been found that 17β-estradiol has a high intrinsic antioxidant activity

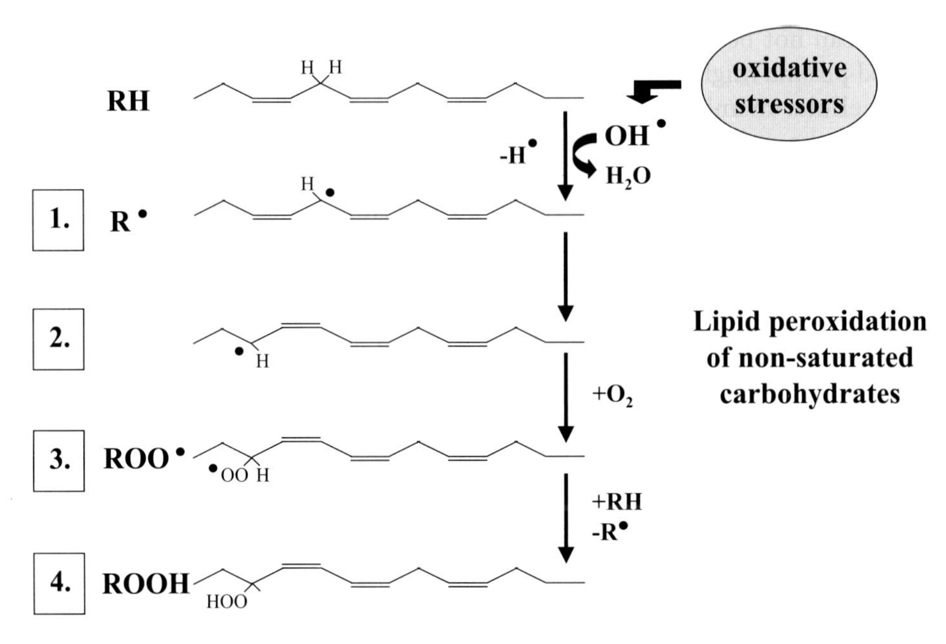

Fig. 23. Lipid peroxidation caused by ROS.
The non-saturated carbohydrate side chains of membrane lipids are the targets for hydroxyl radicals. 1. The first step is hydrogen abstraction. 2. The molecular structure is rearranged. 3. Oxygen is taken up and lipid peroxy radicals (ROO·) are formed. 4. Lipid peroxides (ROOH) are generated

chain of events so that the gross architecture of the membrane is disturbed (Fig. 23). Vitamin E can interrupt this chain of reactions and, therefore, it is called a chain-breaking antioxidant (for review: Traber and Sies, 1996; Halliwell and Gutteridge, 1999). Because of its natural origin α-tocopherol is a very frequently used nutritional supplement and also in use for therapeutical intervention in clinical trials for arteriosclerosis, Parkinson's Disease, and Alzheimer's Disease, admittedly with limited success. One reason for only the limited potency of vitamin E in conditions of oxidative stress in the brain may lie in the limited ability of α-tocopherol to pass the blood-brain-barrier (BBB).

The BBB is considered as a protective barrier for the brain and is formed by glial cells of the brain and by bloodvessels. Although the BBB does not allow free access of molecules from the blood stream transport mechanisms for certain molecules exist and the BBB should not be considered as a stiff impermeable wall. During inflammation and other conditions the function of the BBB is frequently disturbed. There is only a small amount of data available on the specific capacity of phenolic compounds to penetrate the BBB, but it can be assumed that phenolic compounds behave more or less analogously to what is known for some pharmacologic compounds. The major prerequisite of a successful penetration of the endothelial/pericytal cell layer (BBB) is a high lipophilicity. On the other hand these compounds should not be trapped inside the lipid bilayer, which is the case for long-chain alkylated compounds such as vitamin E (Seelig et al., 1994).

Some phenolic compounds are specifically sequestered by the brain, including 17β-estradiol (Pardridge et al., 1980). Selected phenolic compounds, e.g. L-dopa, enter the brain via another system of active transport (Oldendorf, 1971). Nevertheless, influx by diffusion is probably the most important pathway, and molecules bearing multiple hydroxyl groups are basically incapable to enter the brain very effectively. Furthermore, flavonoids tend to be tightly bound to large plasma proteins which lowers the pressure of these molecules to penetrate the BBB. An example is quercetin (Gugler et al., 1975). In summary, it is very likely that estradiol may enter the brain when administered peripherally. The pharmacology and pharmacodynamics is of great importance when considering orally applied estrogen, e.g. during estrogen replacement therapy (ERT), as a drug for the brain. In several animal studies it has been shown that radioactive-labeled estradiol can enter the brain, although the mechanism is not exactly known. Human studies that show that estrogen enters the brain are – to my knowledge – not available but the fact that various menopausal changes that are mediated by the hypothalamus (e.g. hot flushes) are effectively treatable with estrogen strongly argues that estrogen passes the BBB and reaches the brain.

In summary, in this chapter non-genomic, rapid effects of estrogen have been discussed. Many of estrogen's rapid effects concern interactions of estrogen with intracellular signaling pathways (e.g. release of calcium, induction of cAMP formation, activation of MAP kinases). These rapid estrogen effects can not be blocked by inhibitors of transcription and translation strongly suggesting a mode of action independent of ER activation. Since many of these rapid activities are observed at the cellular membrane the discussion concerning the existence of membrane ERs is ongoing. Most of these membrane effects can be explained by estrogen's capacity to interact with membrane-associated functional proteins, including ion channels. Many electrophysiological observations support this view. In addition, estradiol has been introduced as a structural antioxidant that is able to detoxify highly destructive molecules with unpaired electrons, the so-called free radicals, through its phenolic hydroxyl group. This antioxidant activity is independent of ERs and could be of great significance since many neuronal (and non-neuronal) disorders are accompanied or perhaps caused by the generation of free radicals and oxidative stress.

5. General physiological activities of estrogen

Estrogen has multiple physiological functions and is a central modulator at the molecular, cellular, and behavioral level. Of great importance is estrogen's responsibility for sexual differentiation and maturation as well as for the growth of female secondary sexual organs. Estrogen has a variety of effects on the metabolism of macromolecules in both the vagina and the uterus. Among other activities in the mammary gland, estrogens together with progesterone regulate the lactogenesis of the gland. But doubtless, the female sex hormone has multiple functions all over the body due to the fact that ERs are expressed in a variety of tissues. Undisputed is the role of estrogen during sexual differentiation and maturation. There is a great body of literature on the function of estrogen in human sex organs such as in the ovaries and in reproduction as well as on the basic findings of the interaction of estrogens with brain function. Therefore this discussion will end here by giving reference to the literature (for review: James et al., 1976; McEwen et al., 1987; Genazzani et al., 1992; Priest and Pfaff, 1995; Pfaff, 1997; Redmond, 1999). Based on the research of the last two decades, estrogen is also active in the CNS and modifies various neurophysiological processes. Estrogen's main target tissue in the brain with respect to sex-related processes is the hypothalamus. The hypothalmus is an endocrine gland and secretes signal molecules which act on pituitary action (see page 17). It takes a central part in the regulation of other glands including the ovaries and thyroids. Consequently, hypothalamic functions control body temperature, blood sugar, fat metabolism and other central conditions. Estrogen is an important regulator of the hypothalamic-hypophyseal-system and modulates the secretion of luteinizing hormone (LH) and of follicle stimulating hormone (FSH). Of course, estrogen is part of various feedback mechanisms in the hypothalamus and has various activities (for review: Blaustein and Olster, 1989; Baulieu and Kelly, 1990; McCarthy and Pfaus, 1996; Horvath et al., 1997; Herbison, 1998; Frohlich et al., 1999; Etgen et al., 1999). During the development of the mammalian nervous system estrogen affects various processes at different levels (for review: Beyer, 1999).

Lessons from the ERKO-mice

Recalling the initial endocrinological experiments that analyzed the activities of sex hormones: the removal of the gonads in birds prevented reproduction and sexual phenotype as described by Aristoteles. Modern molecular genetics basically use a similar approach to study gene function, the deletion, removal (knock-out) or modification (transgenic approach) of a gene.

On the basis of the available data it appears quite clear that both types of estrogen receptors (ERα and ERβ) fulfill quite distinct physiological roles. Many of these findings are deduced from genetically modified and knock-out mice. First of all it has to be mentioned here that with respect to genetically engineered mice there is a general difference between a *knock-out mouse* and a *transgenic mouse.*

The transgenic mouse carries genetically manipulated alleles of a known gene, in most cases alleles with a disease causing mutation, in its genome. This modified allele has been injected into embryonal stem cells and by recombination procedures during the proliferation of the blastocyst in some cases the allel gets incorporated into the germline leading to the expression of the modified gene in all body cells. A mouse where both copies of one gene have been replaced by the modified allels is called homocygotly transgenic for this novel allel. This technology is now frequently employed in most cases with the hope to create animals with human disease genes that indeed represent features of the human disease and that can be used as models. Such an animal is a worthy model for the investigation of possible pharmaceutical intervention to halt or change the course of the disease process.

With respect to neurodegenerative disorders examples of transgenic mice are various so-called "Alzheimer mice", which imitate some but not all the features of AD. A particular Alzheimer-associated protein, the amyloid β protein (Aβ), overproduced in certain families that develop a familiar form of AD (Games et al., 1995; Masliah et al., 1996; for review: Aguzzi and Raeber, 1998). This overproduction can be caused by mutations in the Aβ-precursor protein (APP). When this mutated APP is overexpressed in a transgenic mouse model, indeed, the overproduction of Aβ also occurs in the animal. These mice are now important tools to test compounds that when given to the animal prevent the overproduction of Aβ. Such a compound could be a potential new anti-Alzheimer drug.

The knock-out mouse on the other hand is different in nature. In a knock-out mouse a particular gene has been removed and if both allels of a gene are removed a homocygous knock-out mouse is created. Such mice are very frequently generated to study the basic physiological function of certain genes. The rational of this approach is that the lack of particular structures and functions, the phenotype, demonstrates the basic function of the knocked out/removed gene. Coming back to the Alzheimer example, indeed various knock-out mice have been produced, e.g. an APP-knock-out. Interestingly, this APP knockout is fully viable and does not really show a dramatic phenotype although it is known that APP may act as neurotrophin and as a extracellular matrix protein at the synapse of neurons. It is clear that all these models underlie manifold mechanisms of compensation. The fact that a knock-out mouse is born and, therefore, survives the lack of the

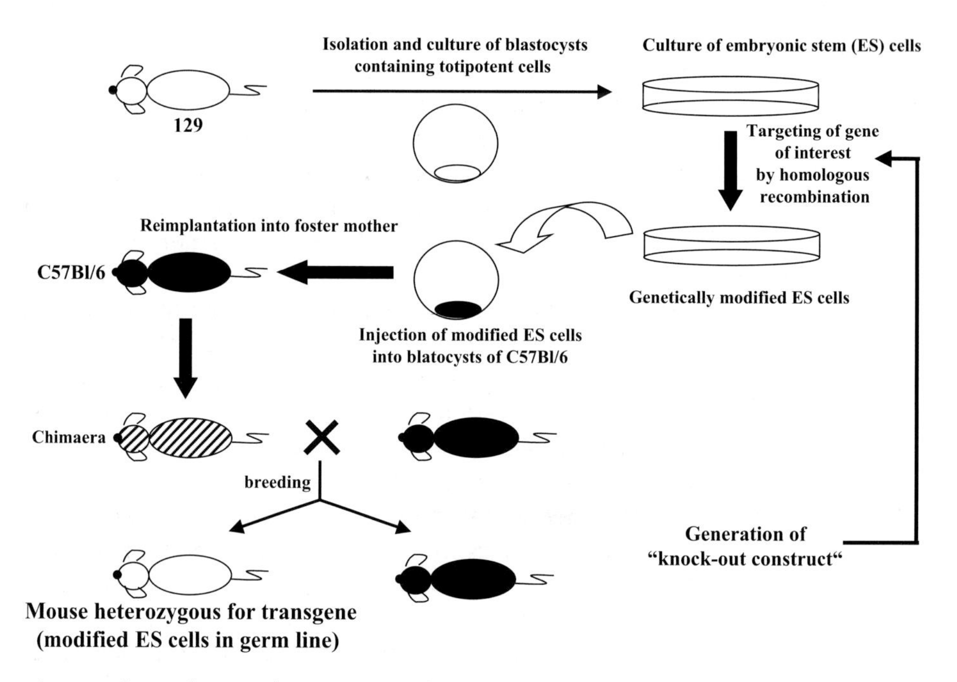

Fig. 24. Scheme showing the generation of a "knock-out" mouse using homologous recombination in embryonic stem cells.
From blastocysts of the mouse strain 129 (which has a brownish fur color), embryonic stem cells (ES cells) are isolated and cultured in a dish. The "knock-out" DNA construct containing the modified gene of interest is transferred into the ES cells, where the gene locus of interest will homologously recombine with the "knock-out" construct. This leads to a targeted genetic modification of the ES cells in the gene of interest. These genetically altered ES cells are injected into blastocysts isolated from the mouse strain C57BL/6, which have a black fur color, and are then implanted into a foster mother, which will give birth to chimeric mice, carrying two populations of cells derived both from modified ES cells and from C67BL/6 blastocysts leading to spots of both black and brownish color. Chimerae are crossed with C57BL/6 mice, which generate brownish offsprings if they are derived from the modified ES cells. At this stage, the mice are heterozygous for the modified gene locus. To generate a homozygous null mutant (a "knock-out"), two heterozygous mice will be crossed

gene of interest already strongly argues for the induction of compensatory and rescue mechanisms. Nevertheless, both approaches, the transgenic mouse and the knock-out mouse, are now very important techniques that have been further developed with very sophisticated protocols so that also tissue-specific knockouts and transgenics can be produced. This is of big importance since if a gene is manipulated (modified or removed) in all body cells the phenotype may be lethal from the beginning preventing any further investigation of this particular model. So far, knock-out approaches have been used to study the function of ERs. The basic experimental strategy for the generation of a complete gene knock-out in mice is simplified shown in Fig. 24.

Of course, ERs are believed to play pivotal roles during development, reproduction, and physiology. With respect to embryonal development it should be mentioned here that both ERα and ERβ are expressed in mice throughout embryogenesis starting around day 10 following conception. Estrogen receptor mRNA was found to be expressed in a whole variety of tissues, including the reproductive tract, the atrial wall, brain, kidney, bladder, neck, and mammary gland (Lemmen et al., 1999). Interestingly, ERβ appeared to be the only ER form to be expressed in the nervous system of the embryo.

In the search for physiological functions of ERs mice homozygous for the disrupted ERα gene (αERKO) (Lubahn et al., 1993; for review: Korach et al. 1996; Couse and Korach, 1999) and ERβ gene (βERKO) (Krege et al., 1998) were created. As demonstrated in Fig. 25 for the ERα a 1.8 kb DNA insert containing the gene for neomycin (NEO) resistance was inserted into special restriction sites (cleavage sites in the DNA) present in exon 2 of the ERα gene which has been subcloned from a so-called genomic library of mouse DNA. This novel gene construct carrying the disrupted ERα gene was targeted into mouse embryonic stem cells and stem cells expressing this gene could be isolated through selection for neomycin resistance (clone selection). Then, the blastocyst was injected into a hyperovulating female mouse to generate chimeric mice carrying the disrupted ERα gene. The correct germ-line transmission and manifestation of the disrupted gene was then confirmed by analytical methods including Southern blotting and PCR analysis. When mice heterozygous for the ERα disruption (ERα–/+) were inbred homozygous ERα–/– could be isolated according to the Mendelian law.

When the αERKO mouse was analyzed it was a great surprise that this animal was viable, considering the central functions of the female sex hormone estrogen during development believed to be mediated by the only known ERα. With todays knowledge about ERβ it is clear that ERβ may compensate ERα's actions in some respects. Technically there was still estrogen binding and specific activity described in the αERKO mouse, which could not be explained. But there were some striking findings in these mice.

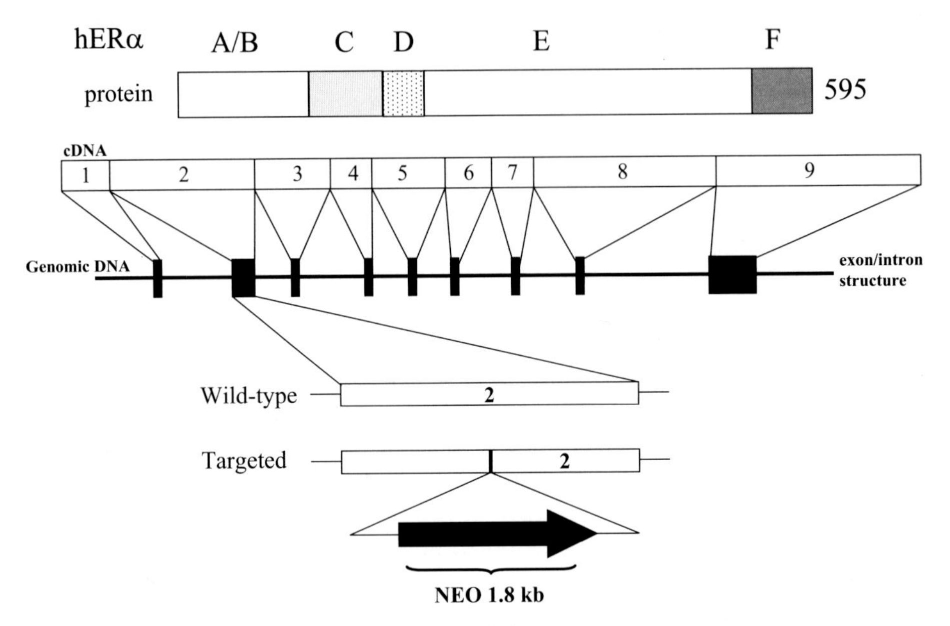

Fig. 25. Vector construction for the generation of the αERKO mouse.
Generation of the αERKO mouse involved the insertion of a 1.8 kb DNA sequence coding for the neomycin resistance gene (NEO) into exon 2 of the gene for ERα (adapted from Couse and Korach, 1999)

The αERKO female mice are infertile and have dysfunctional uteri and ovaries (Lubhan et al., 1993). The male αERKO mice are also infertile with a prominent atrophy of the testes, decreased spermatogenesis and inactive sperm cells. Besides the reproductive tract major pathophysiological changes were the reduced bone density and altered cardiovascular function. Both, in female and male αERKO mice, overall gross phenotypic changes have been reported when compared to the wild-type mice including an altered sex and aggressive behavior. The various findings on αERKO mice are presented in detail in a recent review (for review: Couse and Korach, 1999).

The disruption of the gene for ERβ was performed by targeting exon 3 of the ERβ gene and the generation of homozygous ERβ–/– mice was done similar to the ERα approach (Krege et al., 1998). βERKO female mice developed an arrest in the folliculogenesis and, consequently, infertility. In contrast βERKO male mice are fertile and show a normal sexual behavior (Krege et al., 1998). It has to be mentioned here that also so-called double-k.o. mice (ERα and ERβ-k.o.-mice) are currently created.

Based on the ERKO studies as well as many other observations made since the discovery of a second ER (ERβ) it appears quite clear that ERα and ERβ have quite distinct biological roles in reproductive tissues (for review: Couse and Korach, 1999; Gustafsson, 2000).

6. Estrogen's actions in the brain

When discussing estrogen's activity in the CNS one has to consider that effects observed in *in vitro* cellular models and even more in *in vivo* models are the result of a mixture of actions. It is hard to separate receptor-mediated from receptor-independent activities completely. Various activities of estrogen that depend on or are independent from ERs are intertwined in a complex network of interactions. Nevertheless, there are certain estrogen effects that can be mainly ascribed to estrogen's ability to bind and activate ERs. The receptor-dependency is frequently shown by using ER-antagonists or more recently by employing ERKO-models. But one has to realize that certain antagonists may also have opposing effects at the ER (see page 47/ SERMs). Moreover, k.o.-mice, as just discussed, are also the results of many compensatory mechanisms. However, it is clear, whenever one manipulates an experimental system, the readout of the system itself is affected to some extent.

To understand estrogen's potential activities in the brain first of all the expression of ERs in the brain needs to be highlighted. As a result of these expression it has to be underlined that estrogen has multiple sites of action outside the classical estrogen target region, the hypothalamus. In addition to the ER-mediated classical or so-called delayed effects some of estrogen's rapid and non-genomic activities in the CNS are highlighted. All these activities have to be considered when, ultimately, investigating neuroprotection mediated by estrogen as will be pointed out later. But since many of estrogen's activities in the brain are of genomic nature the next paragraph gives some information concerning the presence of the mediators of this genomic activity in the brain, the ERs.

Estrogen receptors in the brain

The temporal and spatial expression of ERα protein is well described (Shughrue et al., 1997; for review: Merchenthaler and Shughrue, 1999; Beyer, 1999). However, only recently have various studies been launched investigating the expression of ERβ. Although, in situ hybridizations have been performed to detect the intracellular presence of ERβ-mRNA due to the lack of good and specific ERβ-antibodies the distribution of ERβ protein remains to be revealed in detail. Interestingly, initial studies suggest a subtle overlap in the expression of ERα and ERβ mRNA in various regions of the developing and adult forebrain such as in the cerbral cortex and hippocampus (Li et al., 1997; Osterlund et al., 1998; Shughrue et al., 1997). Indeed, a detailed analysis of ERβ mRNA expression clearly revealed that this type of receptor has a much wider occurrence than ERα mRNA. In detail, while the ventromedial nucleus of the hypothalamus, the subfornical organ, and the dentate nucleus of the cerebellum appear to express exclusively ERα mRNA all other brain regions that show the presence of ERα mRNA also express ERβ mRNA. Very prominent ERβ expression is reported for the neocortex, the enthorhinal cortex, and the Purkinje cell layer of the cerebellum. These regions are all characterized by very low expression of ERα as shown by Shughrue and coworkers (Shughrue et al., 1997). The same study demonstrated that a much greater ERβ mRNA expression is found in the dorsal hippocampus compared to ERα mRNA. In some areas of the brain the co-expression of both types of receptors is suggested which may have significant implications for the estrogen target gene transcription. Therefore, not only homodimers of ERα and ERβ but also receptor heterodimers ERα/ERβ may occur. Receptor heterodimers may target other genes on the nuclear DNA than ER-homodimers (ERα/ERα and ERβ/ERβ). This possible differential induction of estrogen target genes is an important topic of the author's research.

The detection of ERβ-mRNA in brain regions that are responsible for cognitive functions, learning and memory revealed new insights into putative activities of estrogen in brain areas outside the hypothalamus. One direct consequence is that estrogen may modulate the function of cortical and hippocampal neurons via estrogen receptor-dependent processes. The hippocampus is highly sensitive for neurodegenerative challenges and is the prime target in chronic neurodegenerative disorders such as Alzheimer's Disease or during immediate insults as occurring in stroke.

Neuroactivities of estrogens in brain areas outside the hypothalamus: the effect of sex differences

Sex differences are well-known in psychiatric disorders. Examples are depression, schrizophrenia or the sensitivity for pain. Moreover, also the incidence and progression and the response to treatment of movement disorders, affective state etc. appears to be dependent on gender (Kimura, 1992; Regier et al., 1988; for review: Casper, 1998; McEwen et al., 1998). Moreover, there is an important link between estrogen levels and memory function and the occurrence and incidence of dementia, which will be pointed out later (Phillips and Sherwin, 1992). Interestingly, it is has been also shown that estrogen takes part in the regulation of the neurogenesis in the dentate gyrus, the part of the hippocampus formation that continues to produce new neurons in the adult brain (Tanapat et al., 1999).

The activities related to sexual behavior of female rats of secreted ovarian hormones on neurons of the ventromedial hypothalamus are: control of neuropeptide gene expression, regulation of second messengers, induction of oxytocin receptors, progesterone receptors, and the control of cyclic synaptogenesis (for review: Harlan, 1988; Kow et al., 1994; McEwen BS et al., 1997; Frankfurt et al., 1990).

The hypothalamus is the traditional site for the investigation of the effect of estrogen and estrogen receptors with respect to the control of reproductive and sexual behavior. But the extremely great diversity of estrogen's neuroactivities suggests that brain areas outside the hypothalamus are also important targets for estrogen. As mentioned, estrogen receptors are widely distributed throughout the brain raising the question for (1) extrahypothalamic target tissue in the brain and (2) for estrogen's effects on processes not involved in sexual maturation, differentiation, and sexual behavior. Of course, the hypothalamus is not only the brain region involved in the control of sex-related functions, it affects many different neuroendocrine events directly and indirectly. The main extrahypothalamic brain regions that are affected by estrogens (and progestins) are (for review: McEwen and Alves, 1999):

- the cholinergic system,
- the serotonergic system,
- catecholaminergic neurons,
- the spinal cord,
- the hippocampus,
- glia, endothelial cells, and the blood-brain barrier.

Effects of estrogen on the cholinergic system

Acetylcholine (ACh) is a neurotransmitter of central importance for cognitive functions. Cholinergic neurons project from the basal forebrain to the cerebral cortex and hippocampus and play a central role in learning and memory. Estrogen is thought to enhance cognitive functions by modulating the production of acetylcholine in basal forebrain neurons. Animal studies performed with rats demonstrated a direct trophic effect of estrogen on cholinergic neurons including the induction of choline acetyltransferase (ChAT), the main enzyme promoting the formation of ACh in the synapse, and of acetycholinesterase (ACh-esterase), the enzyme that mediates the chemical breakdown of ACh in the synaptic cleft (for review: Gibbs and Aggarwal, 1998; Gibbs, 2000) (Figs. 26 and 27).

The mechanism of the interaction between estrogen and the cholinergic

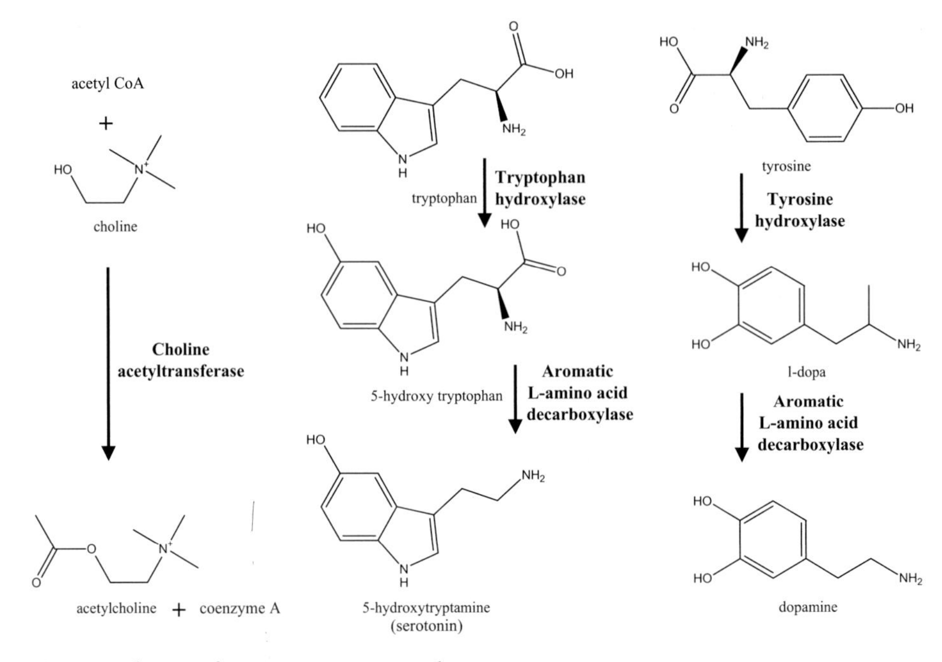

Fig. 26. Pathways of neurotransmitter synthesis

system is not exactly known but it is intriguing to believe that NGF may be the functional intermediate and modulator. Indeed, the interaction of NGF with the cholinergic system is well described. ERα colocalizes with the low-affinity NGF receptor in cholinergic neurons of the basal forebrain, at least in newborn rats (Toran-Allerand et al., 1992). With the elucidation of a possible direct "cross talk" between estrogen and neurotrophins, this assumption makes a lot of sense (for review: Toran-Allerand et al., 1999). In a recent study investigating the estrogen binding and expression of ER in the rat basal forebrain it was found that ERα is the predominant ER in cholinergic neurons while only a few cells contain ERβ (Shughrue et al., 2000). Activation of ERα by estrogen in the basal forebrain cholinergic neurons of the adult rat brain may directly modulate the NGF/NGF-receptor system. This further supports the view that estrogen directly modulates the activity of cholinergic neurons in rats and may in the future provide detailed insights into how estrogen improves cognitive functions in women. Moreover, these findings may explain the cognition-enhancing effects of estrogen in women, which will be pointed out later. Also it could be relevant for the observation that estrogen improves the effect of inhibitors of the acetylcholinesterase inhibitor when used for Alzheimer treatment; estrogen could support the ACh-system (Schneider and Farlow, 1997).

Postmortem studies employing the brain of AD patients have found a signifcant reduction in the presynaptic markers for cholinergic neurons in

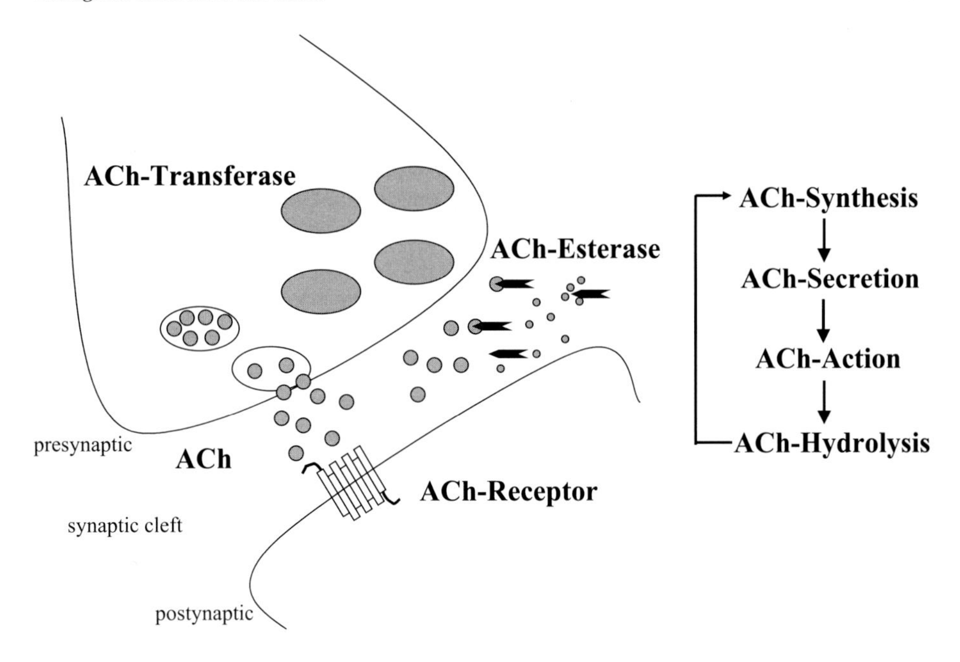

Fig. 27. Turnover of acetylcholine.
The life cycle of the neurotransmitter ACh is shown. ACh is released from synaptic vesicles into the snyaptic cleft and acts at its ACh-receptors at the postsynaptic site. ACh-esterase enzymatically degrades ACh in the synaptic cleft

the cortex of AD patients and in 1982 the *Acetylcholin-Deficiency-Hypothesis* has been formulated. In fact more than 75% of neurons of the nucleus basalis of Meynert in the brain of AD patients are degenerated, which is the reason for the cholinergic deficit in AD (Whitehouse et al., 1982). Since then this hypothesis has been challenged, revised and confirmed in various follow-up investigations *in vitro* and *in vivo* but is still considered for the design of cinical studies that target the acetylcholine system for AD therapy (for review: Cummings and Back, 1998; Schneider, 1998).

It is clear that the cholinergic deficit is one major alteration during AD pathogenesis and, obviously, accounts for the detrimental changes in memory and cognition. A recent study aimed to determine whether higher brain levels of choline acetyltransferase (ChAT), the central enzyme in the synthesis of the neurotransmitter acetylcholine, are indeed associated with improved neuropsychological functions in patients with AD. In this study also tissue biopsies have been included and it was found that there is a significant positive correlation between the level of ChAT and certain neuropsychological test scores, including the frequently used so-called "Mini-Mental State Examination". This study clearly indicates that the degeneration of the cholinergic system *in vivo* correlates with decreasing cognitive function in patients with AD (Baskin et al., 1999). This finding is in great support of the acetylcholine deficiency hypothesis of AD. Recently, a benefical effect of estradiol on ACh-mediated memory processing in mice was reported (Farr et al., 2000).

The interaction between estrogen, the cholinergic system, and aging has also been investigated. It is presented that in the cholinergic system, estrogen mediates the number of cholinergic receptors, cholinergic neurons, and cholinergic-modulated memory systems in both young and old animals. Moreover, independent of age, physiological levels of ovarian steroids are beneficial to the overall neuroendocrine functions in the CNS, and long-term deprivation from ovarian hormones such as through ovariectomy is deleterious. It is concluded that the loss of ovarian steroid hormones in the female is a high impact factor in the health of the functions of the CNS and in age-related changes (for review: Miller et al., 1998).

Effects of estrogen on the serotonergic and catecholaminergic system

Serotonin (5-hydroxytryptamine) is an indolamine (Fig. 26). Large areas of the brain are innervated by serotonergic fibers. The most prominent serotonergic pathway emerges from a brain structure called dorsal raphe nuclei, which is known to be involved in sleep control. The serotonergic system is also a major target for antidepressant, such as serotonin reuptake inhibitors. The serotonergic system is involved in the regulation of a variety of body functions including reproduction, mood, sleep, and cognition. A lot of evidence suggests that this serotonergic activity is regulated by ovarian steroid hormones.

The strongest argument for a potent role of sex hormones in any kind of physiological process or disease state is the presence of sex differences in the expression or function of the particular transmitter system. Such sex difference are indeed observed. Compared with male rats, female rats have higher serotonin levels in certain brain areas including the hippocampus (Kawakami et al., 1978; Dickinson and Curzon, 1986; Haleem et al., 1990). Similarly, the overall turnover of serotonin which indicates serotonergic activity is also increased in female rats and the level and activity of serotonin are changing with variations in the estrogen levels as occurring during the estrous cycle or in pregnancy (Kueng et al., 1976; Biegon et al., 1980). These findings are supported by types of experiments that are referred to in this book very frequently which is the use of ovarectomized rats studied in comparison with those supplemented with estrogen. Estrogen treatment enhances the activity of the serotonergic system in the female rat brain (Luine et al., 1975; Munaro, 1978; Biegon et al., 1983; Cone et al., 1981; Chomicka, 1986). Whether estrogen has a more direct or indirect effect on serotonergic neurons is not clear so far. Recently, the presence of estrogen-inducible progesterone receptors in many serotonin neurons and in the non-serotonergic cells in a brain structure called dorsal and ventral raphe was found in the rhesus macaques (Bethea, 1993; for review: Bethea et al., 2000). The presence of progesterone receptors in these neurons may explain estrogen's effects on the serotonergic neurons since progesterone receptor expression is influenced by estrogen. In further support of a role of ovarian hormones in serotonergic neurons are the findings that ovarian steroids increase the expression of the key enzyme of serotonin biosynthesis, the tryptophan hydroxylase, and decrease the expression of the serotonin transporter. Very recently, these results were confirmed by the same group and it was found that the stimulatory effect of estrogen (and progesterone) on tryptophan hydroxylase protein in the dorsal raphe of macaques correlates with the observed effect at the level of mRNA expression (for review: Pecins-Thompson et al., 1996, 1998; Bethea et al., 2000).

No ERα expression is observed in serotonergic neurons but there is expression of these receptors in neurons adjacent to the serotonin cells, which would open at least the possibility of some regulation of the serotonergic neurons through estrogen. But again, the discovery of the second ER, ERβ, may change the whole picture since, indeed, ERβ mRNA has been found in the dorsal raphe of the rat (Shughrue et al., 1997). Future studies may reveal if the estrogen effects on the serotonergic system is mediated by binding to and activation of ERβ.

The knowledge about the important function of the serotonergic system in the mood status and in cognition raises many interesting questions with respect to the involvement of estrogen in these processes. Depression is known to be more common in women than in men and the incidence of depression appears increased at times of changing hormone levels in women. Antidepressant effects in humans have been reported for high dosages of estrogens (Klaiber et al., 1996) and estrogen may also improve the responsiveness of women to antidepressant drugs. The latter has been reported based on a clinical trial of fluoexitin (Prozac, EliLilly, Indianapolis, IN; Schneider et al., 1997). Fluoxetine is a selective serotonin reuptake inhibitor and therefore indirectly increases the presence of serotonin in the synaptic cleft. This is believed to mediate its anti-depressant effect according to the serotonin hypothesis of depression, which claims that decreased levels of 5-HT in the synaptic cleft as one cause for the depression state (for review: Mann, 1999). A closer look into the current literature on this topic reveals that estrogen therapy may indeed be useful for the treatment of depression and mood changes in peri- and postmenopausal women (for review: Archer, 1999). But estrogen modifies not only serotonergic activity and, of course, also influences the activity of several other neurotransmitters that are associated with antidepressant effects.

In addition to the cholinergic and serotonergic system, estrogen has effects also on catecholaminergic neurons including noradrenergic and dopaminergic neurons.

Catecholaminergic neurons in the brainstem express a small number of ERs and the level of tyrosine hydroxylase mRNA is influenced by estrogens (Heritage et al., 1980; Liaw et al., 1992). Sex differences have been described for the dopaminergic system in the striatum. Estrogen depletion of animals by ovarectomy decreases and estrogen supplementation enhances the release of dopamine (Di Paolo et al., 1985; Becker and Beer, 1986; Becker, 1990). Despite these pro-dopaminergic effects of physiological levels of estrogen, no ER expression has been observed in the striatum and these estrogen activities have been ascribed to non-classical estrogen effects, probably on the neuronal membranes (Xiao and Becker, 1998; for review: Rupprecht and Holsboer, 1999). With respect to Parkinson's Disease (PD) pharmacologically high estrogen levels as reached through oral contraceptives or estrogen replacement therapy, may exacerbate the disease symptoms and are, therefore, described as anti-dopaminergic in nature (Riddoch et al., 1971; Barber et al., 1976; Bedard et al., 1977).

The results of recent clinical trials investigating the effect of estrogens on PD challenge these earlier findings since in these investigations rather positive or no effects of estrogen were found (Strijks et al., 1999; for review: Kompoliti, 1999). Further extrahypothalamic brain areas outside the hypothalamus that are affected by estrogen activities also include the hippocampus which has been mentioned earlier as one main site of estrogen action and will be discussed with respect to gender differences below. Also the spinal cord is a principal target of estrogen action and ERα and ERβ mRNA is detectable throughout the rostral-caudal extent of the brain and spinal cord (Shughrue et al., 1997). Compared to other regions of the nervous system the expression of ER in the spinal cord is very limited (Hosli and Hosli, 1999). In favor of estrogen actions in the spinal cord is the finding of antinociceptive and analgesic activities of estrogens. Indeed, there is a striking gender difference in the sensitivity to pain (for review: Mogil, 1999).

Activities of estrogen on glial cells

Of major importance for proper functions of the CNS are also the glial cells of the brain and the endothelial cells that comprise the BBB. Astrocytes (astroglia) are the most numerous glial cells in the brain and fulfil many regulatory functions. These astroglia cells are also clear target cells for estrogen as well as for testosterone. Sex hormones influence the level of the expression of glial fibrillary acidic protein and the growth of astrocytic processes.

Many effects are observed in the hypothalamus and therefore it is suggested that astrocytes may take part in the development of sex differences in synaptic connectivity and in estrogen-induced synaptic plasticity in the adult brain. Since astrocytes produce certain growth factors (e.g. Insulin-like-growth-factor, IGF-1) and inflammatory mediators, estrogen may influence these astrocytic activites as well. Interestingly, the expression of aromatase, the enzyme that produces estrogen from testosterone, is induced de novo in astrocytes in lesioned brain areas of adult male and female rodents, which may have an impact on regenerative functions of estrogen following neuronal insults (for review: Garcia-Segura et al., 1999). Recently, evidence has been presented for the presence of ERβ in rat microglial cells (for review: Mor et al., 1999). In contrast to astroglia, microglial cells are much smaller in size. Microglial cells are able to migrate through neuronal tissue and in its activated form are important mediators of inflammatory reactions. Therefore, estrogen activities have to be taken into account also during typical inflammatory and non-inflammatory functions of microglial cells. I will come back to this point later also with respect to neurodegeneration since in Alzheimer's Disease inflammatory reactions mediated by microglia may be of central importance for the disease process (for review: McGeer and McGeer, 1999; Halliday et al., 2000). The expression of ERs is also observed in myelin-generating glial cells, such as the Schwann cells (in the periphery) and oligodendrocytes (in the brain), and may have an effect on the pathogenetic events occurring during demyelinating disorders such as Multiple Sclerosis (for review: Gudino-Cabrera and Nieto-Sampedro, 1999; Jung-Testas and Baulieu, 1999). The exact activities of estrogen in these glial cells, in particular *in vivo*, are under investigation. But studies *in vitro* and *in vivo* show a direct genomic effect of estrogen (Langub and Watson, 1992; Santagati et al., 1994) since it can influence their gene expression patterns. One important gene that appears to be regulated by estrogen is the gene for apolipoprotein E (Stone et al., 1997), which plays a major role in cholesterol transport. In mice deficient in apolipoprotein E, estrogen has no effect on synaptic sprouting in the hippocampus (Stone et al., 1997) suggesting that it is this particular interplay that takes part in the sprouting process.

In endothelial cells of the BBB estrogen enhances the activitiy of glucose transporter-1 (Shi and Simpkins, 1997) and ERα-expression has been shown in cerebrovascular endothelial cells (Langub and Watson, 1992). This may be the basis for the role of ovarian steroids in the glucose-metabolism in the female brain. It is known that ovarectomized rats exert a decreased capacity for glucose utilization and, on the other hand, estrogen treatment increases this capacity (Namba and Sokoloff, 1984; Bishop and Simpkins, 1995). Consistently with the enhanced glucose utilization in animals in postmenopausal women experiencing an estrogen replacement the overal cerebral blood flow during memory tasks was found to be significantly increased (Resnick et al., 1998).

Are there gender differences in brain function?

Yes there are. The male and female brain is different with respect to structure and function. There are variations in nerve cell number, brain morphology, and neuronal connectivity. These variations are the substrates for differences in brain physiology, cognitive development, and behavior. Gender differences are believed to be mediated by the activity of the gonadal hormones during development and also later in life (for review: Kimura, 1992).

Sexual differentiation and gross gender differences in brain structure and function

Early in development the gonads as well as the nervous system are not differentiated and neurons from both sexes contain ERs. A central step of male sex differentiation is the secretion of testosterone by the differentiated male gonads, the testes. Absence of testosterone during the development leads to the development of gonads and brain structures with the female phenotype, despite the male genotype. During female sex differentiation the secretion products of the ovaries, mainly estradiol, regulates the female sex phenotype also in the brain. So in both basic pathways of sexual differentiation of the brain estradiol plays the key role. In female estradiol is present throughout the development and adulthood, in male testosterone is converted to estradiol. Detailed reviews on the function of hormones on the sexual differentiation of the nervous system and the impact on behavior are available (for review: MacLusky and Naftolin, 1981; Breedlove, 1994). But what are the actual gender differences in the brain?

At the time of birth, an obvious gender difference is an increased brain weight and volume in males. Besides rather gross variations, including brain weight and volume, sex-related differences exist also in the magnitude of the asymmetry of the brain's hemispheres. It is understood that sex-specific genetic factors as well as non-genetic and non-gender-related factors can alter brain development and the extent and pattern of cortical asymmetry, in general (Diamond, 1991). Interestingly, an enhanced hemispheric connectivity in women is reported which could be caused by a large corpus callosum, the fiber tract that connects the left and the right brain hemisphere. Brain structure is the substrate for brain function. Due to the quite obvious gender differences in brain anatomy, many studies focussed on differences in brain function. In general, there is a higher cortical blood flow in women compared to men and variations in the cerebral metabolic activity in certain regions (Gur et al., 1982; 1995). Moreover, the distribution of some neurotransmitters have been also reported to differ (Heller, 1993). Variations in the so-called lateralization of brain functions can be observed and have been investigated in various psychological tests that analyze visual-spatial and verbal skills. Some of these functional differences in brain and the hormonal influences on the sexually differentiated brain are summarized in a recent review (for review: Casper, 1998). For the present discussion one should keep in mind that the brain of women and men are indeed quite different. The complexity of the brain does not allow interpretations when observing gross differences. Since the topic in this book is the collection of current knowledge on the interplay between estrogen and nerve cell death and neuroprotection the following chapter focuses on the brain area that is one main target of neurodegeneration, the hippocampus.

Sex differences in the function of the hippocampus

For approximately 50 years the hippocampus is known as the brain area important for long term memory storage in mammals including humans. The hippocampus is essential for initial storage of long-term memory for a period of days to weeks before the memory is established in other brain regions. Various well defined synaptic pathways in the hippocampus are known to mediate long-term synaptic plasticity and the establishment of long-term potentiation (LTP; see below). Since the hippocampus is the area of the brain involved in learning, memory, and cognition, structural and functional gender differences are of particular interest. Indeed, there are sex differences in the hippocampal function such as in hippocampal LTP (Maren et al., 1994). Changes in the polarization of neuronal membranes are normally very rapid events. The depolarization of the membrane is the basis for the generation of an action potential which is the biological substrate for the transmission of electric signals along neuronal membranes. LTP means long term potentiation and is postulated as one cellular mechanism of memory storage in the mammalian brain and is based on long-lasting electrophysiological alterations in the neuron and is, therefore, an enduring form of synaptic plasticity. LTP was discovered in the hippocampus of rabbits by Timothy Bliss and Terje Lomo in 1973. Other investigators also demonstrated LTP in hippocampal slices prepared from rodents. Brain slices can be prepared from various brain regions. Organotypic brain slices of various thickness (approximately 400 nm) are prepared with a microtome and can be then cultured up to weeks in tissue chambers filled with culture medium. The big advantage of such organotypic slices, which are also frequently employed in *in vitro* toxicity paradigms, is that the tissue organization, here the intrinsic hippocampal circuitry and the interactive network between neuronal and glial cells is largely intact compared to single cell preparation also frequently used for electrophysiological studies. Compared to single cell cultures slice preparations are much closer to *in vivo* conditions.

LTP resembles memory in several way since it can be induced very rapidly (within seconds) and it may last for days or weeks (Bliss and Lomo, 1973; Bliss and Gardner-Medwin, 1973; for review: Miller and Mayford, 1999; Knierim, 2000; Bennett, 2000). In the hippocampus there are three principal types of neuronal cells that are involved in the excitatory cicuitry: the granule cells of the dentate gyrus (DG), pyramidal cells of the region CA3, and pyramidal cells of the region CA1. The dentate gyrus receives projections from the entorhinal cortex and the dentate gyrus cells project so-called mossy fibers to the CA3 region. The CA3 pyramidal cells then project to the CA1 pyramidal cells (Fig. 28).

Estradiol enhances LTP at synapses of neurons in the CA1 region. Considering LTP as one process establishing memory estrogen appears to consolidate this memory, an important finding. This estradiol effect can be observed during the estrous cycle but also following exogenous treatment. Employing *in vitro* slice preparations it was also observed that estrogen increases the overall excitability of hippocampal CA1 pyramidal cells. Therefore, estrogen has profound effects on the hippocampal circuits and on hippocampal function including LTP (for review: Woolley, 1999).

In addition, certain behavioral tasks that are assigned to hippocampal function are solved differently by male and female animals (Roof and

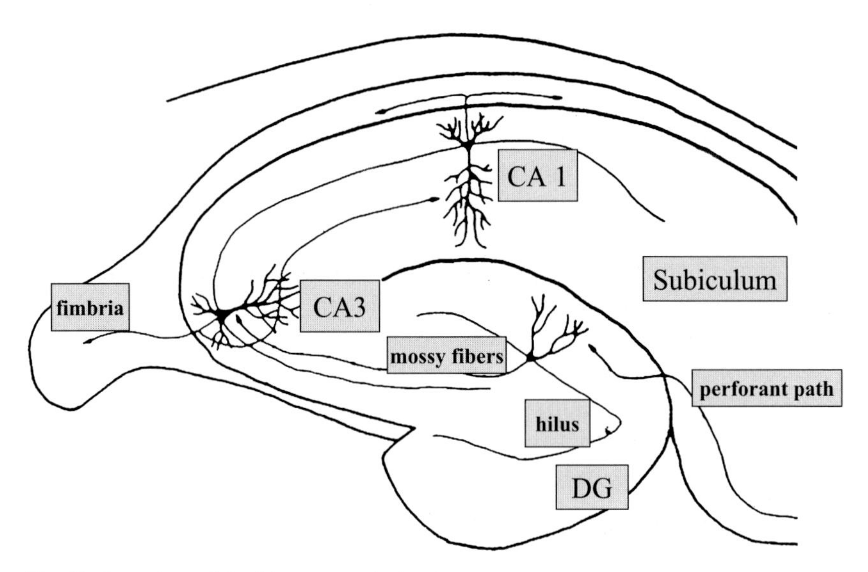

Fig. 28. Overview of the hippocampal circuitry.
(Adapted from Woolley, 1999)

Havens, 1992; Roof et al., 1993; Galea et al., 1996; Kavaliers et al., 1996). Some of these variations may be directly caused by differences in the hippocampal structure that have been reported. The dentate gyrus structure of the hippocampus of males has an increased total number of granule neurons compared to females and similarly the hilus structure of males has more mossy fiber synapses (mossy fibers are the axons of cells in the dentate gyrus). On the other hand, females have a greater number of such mossy fiber synapses in the CA3 region compared to males. Again, the levels of circulating gonadal hormones during development and in the adult can be accounted for these differences. It is clear that during the development of the nervous system the volume of CA1 and CA3 regions of the hippocampus is influenced by estrogen treatment (Cordoba-Montoya and Carrer, 1997; Daniel et al., 1997; Packard and Teather, 1997a, b; Luine et al., 1998; Isgor and Sengelaub, 1998; for review: Arnold and Breedlove, 1985; McEwen et al., 1995).

Briefly, there is an ongoing influence of gonadal hormones on the function of the nervous system also in the adult. Consequently, with respect to disease sex-related variations may also influence the prevalence and the pathogenesis as well as the response to treatment of psychiatric disorders. Although it is known that e.g. more women than male suffer from age-associated Alzheimer's Disease, a fact that is very likely linked to the massive hormonal changes (drop in estrogen levels) after menopause, sophisticated methods of the epidemiology are necessary for a clear picture of the relation of gender and disease. On the basis of rather large-scale epidemiologic studies it was revealed that, indeed, there are major gender differences in psychiatric disorders, including major depression, alcohol dependence, anxiety disorders, or schizophrenia (for review: Casper, 1998).

In summary, besides dramatic activities during the development of the CNS also during adulthood estrogen has tremendous effects on (1) the number of synapses in hippocampal neurons, (2) the hippocampal dendritic spine density, and (3) the hippocampal LTP and hippocampal learning, in general. The various effects of estrogen on structural and functional synaptic plasticity in the female rat hippocampus have been reviewed (for review: Woolley, 1999).

"Non-classical" activities of estrogen in the brain

Estrogen as "neuroactive steroid" and estradiol's non-genomic effects at neuronal membranes

Following this short detour about the gender differences for instance in hippocampal structure and function now the rapid, non-genomic, and therefore "non-classical" effects of estrogen in the brain and possible consequences will be discussed. As already mentioned at various points in this book steroids can function not only through the activation of intracellular steroid receptors but rather may also act locally at the membrane or inside the cell. This membrane activity has been ascribed, at least in part, to the existence of membrane-binding sites for steroids or even to membrane receptors (see page 49). Many experimental data on the non-genomic interaction of steroids with neuronal membranes has led to the term "neuroactive" steroids.

In 1992 Paul and Purdy summarized potential neuroactivities of various steroids and they concluded that "neuroactive steroids are natural or synthetic steroids that rapidly alter the excitability of neurons by binding to membrane-bound receptors such as those for inhibitory and (or) excitatory neurotransmitters" (for review: Paul and Purdy, 1992). Therefore, neuroactive steroids are capable of binding to and modulating the function of membrane receptors for neurotransmitters meaning that steroids find no classical ERs that are localized at or in the neuronal membrane but rather find "binding sites" at well-known transmitter receptors. This indirect modulatory activity of neurosteroids introduced a complete new quality of non-genomic steroid action in the brain. The fact that steroids and especially estrogen are present all over the body suggests basic actions also in the brain. The fact that estrogen belongs to the family of "neuroactive steroids" revisited the way of thinking about sex hormones and their actions. Again it has to be mentioned here that of course it was known for long that sex hormones act on the hypothalamus, very likely via specifc ERs. But now it became evident that there is a whole range of interactivities with neuronal membranes independent of ERs all over the CNS.

Initially, the best-studied neuroactive steroids were various 3 alpha-hydroxy ring A-reduced pregnane steroids including the major metabolites of progesterone and deoxycorticosterone, 3 alpha-hydroxy-5 alpha-pregnan-20-one (allopregnanolone) and 3 alpha, 21-dihydroxy-5 alpha-pregnan-20-one (allotetrahydroDOC), respectively (for review: Paul and Purdy, 1992). These compounds have been shown to act as allosteric modulators of specific neurotransmitter receptors, which means that they indirectly influence neurotransmitter receptor function. But in recent years this basic view of neuroactive steroids has been challenged because (1) binding sites have been identified for a whole variety of classical steroid hormones including progesterone, glucocorticoids, aldosterone, testosterone and estradiol and (2) those steroids can affect also various intracellular signaling pathways, which, ultimately, may change gene transcription. So the initial view that the activity of neuroactive steroids is restricted to the modulation of the function of membrane receptors and are strictly non-genomic has changed. Intracellular ERs could also be involved. Indeed, additional genomic activities may come from the activation of intracellular steroid receptors after the intracellular conversion of neuroactive steroids to receptor-active metabolites. The latter has been shown for 3α, 5α-THP and 3α, 5α-THDOC that are both active at the receptor for GABA (γ-amino-butyric acid), the most prominent inhibitory neurotransmitter in the brain. This was a very striking finding, too. A steroid that does not act at a receptor can be converted inside the cell into a receptor-activating steroid. Initially it was believed the 3 alpha-hydroxysteroids do not interact with classical intracellular steroid receptors but rather bind stereoselectively and with high affinity to GABA-receptors. Biochemical and electrophysiological studies have shown that these steroids markedly augment GABA-activated chloride ion currents in a manner similar (but not identical) to that of anesthetic barbiturates. But after intracellular oxidation of 3α, 5α-THP into 5α-DHP and 3α, 5α-THDOC into 5α-DHDOC, the oxidation metabolites are ligands for the progesterone receptor. Moreover, today it is known that in addition to the GABA-receptor also other neurotransmitter receptors are targets of neuroactive steroids (for review: Paul and Purdy, 1992; Rupprecht and Holsboer, 1999). Therefore, as it can be seen from these few observations, the initial concept of neuroactive steroids as modulators of neurotransmitter systems at neuronal membranes is completely revised. But how did the concept of "neuroactive steroids" as rapid actors at the neuronal membrane evolve?

Majewska and coworkers in 1986 showed that the steroids 3α, 5α-THP and 3α, 5α-THDOC modulate the neuronal excitability via their interaction with $GABA_A$ receptors (Majewska et al., 1986). The neuroactive steroids were able to enhance the GABA-elicited chloride-current of this ligand-gated ion channel. By definition ligand-gated ion channels are transmembrane proteins which can be comprised of different protein subunits forming transmembrane pores and which change their permeability for certain ions upon binding of the ligand to the ligand binding site at the extracellular site of the channel protein. From today's standpoint it is believed that 3α, 5α-THP and 3α, 5α-THDOC are so-called positive allosteric modulators of $GABA_A$ receptors. Allosteric modulation means that the interaction with the receptor occurs not at the ligand binding site but at the spatially distinct allosteric site. They increase the frequency or the duration of the openings of this channel and more chloride ions pass the pore complex. Given the importance of the $GABA_A$ channel for the membrane potential, the excitability of nerve cells and, therefore, for the transduction of an electrophysiological signal, the enhancement of this receptor signal by these neuroactive steroids may have a major impact. Other neuroactive steroids such as dehydroepiandrosterone sulfate (DHEA-S) and pregnenolone sulfate are described as antagonists of the $GABA_A$ receptor (for review: Paul and Purdy, 1992; Lambert et al., 1995; Rupprecht and Holsboer, 1999).

But as mentioned above besides the $GABA_A$ receptor also other neurotransmitter receptors are modulated by the action of steroids, including the receptors for glycine, acetylcholine, and 5-HT_3 (serotonine). Of specific interest especially when discussing potential effects of the modulation of neurotransmitter receptors in excitotoxicity, is that steroids have also been shown to modulate the function of the N-methy-D-aspartate (NMDA) and the amino-hydro-methyl-isoazol-propion acid (AMPA) type of glutamate receptor. This is of vital importance for the cell since excitotoxicity means that the activation of neurotransmitter receptors (e.g. NMDA-receptor) by high concentrations of ligand may lead to cell death. Therefore, the interaction of estrogens with excitatory transmitter receptors, in general, are of central importance. And finally, the modulatory potential of neuroactive steroids is not restricted to allosteric interactions with ligand-gated ion-channels since a "cross-talk" of steroids is also possible with receptors that are coupled to intracellular G-protein signaling including the receptor for oxytocin (Gu and Moss, 1996; Grazzini et al., 1998).

Possible interactions of some neuroactive steroids with various neurotransmitter receptors have been summarized recently (for review: Rupprecht und Holsboer, 1999). A lot of interest has been created concerning progesterone and membrane activities. But also the sex hormones 17β-estradiol and testosterone interacts with neuronal membrane receptors. Estradiol is reported as a negative modulator of the 5-HT_3 and the NMDA receptor and as a positive modulator of kainate receptors (Gu and Moss, 1996; Weaver et al., 1997; Wetzel et al., 1998). Testosterone negatively influences the 5-HT_3 receptor action. With respect to the mechanism, most experimental data suggest that there are no specific binding sites for steroids at the receptors but rather that due to their lipophilicity the steroids insert into the membrane at the receptor-membrane interface. This insertion changes the membrane characteristics, such as fluidity, and modulates the function of the neurotransmitter receptor allosterically. An allosteric modulation of neurotransmitter receptors at the neuronal membrane depends on the particular chemical structure of the neuroactive steroids, on the particular amino acid composition of the integral membrane proteins, and, of course, on the general constituents of the neuronal membrane, which determine their physico-chemical properties. The picture has further developed since the neuroactivity is not restricted to the membrane site and also direct genomic effects may occur. As mentioned previously, neuroactive steroids such as 3α, 5α-THP and 3α, 5α-THDOC can not bind and activate intracellular steroid receptors (Gee et al., 1988) but, surprisingly, chemical metabolites of these compounds are ligands for classical steroid receptors (Rupprecht et al., 1993). Following the intracellular oxidation into 5α-DHP and 5α-DHDOC progesterone receptor-dependent gene transcription is activated. And indeed, while 3α, 5α-THP and 3α, 5α-THDOC are not ligands for the progesterone receptor, their oxidation products are. Through this example, the overlap of classical genomic and non-classical non-genomic activities of steroids becomes clearly visible. Addition of 3α, 5α-THP and 3α, 5α-THDOC to neuronal cell systems modulate $GABA_A$ receptor action at the membrane. But due to the lipophilicity these molecules enter the intracellular compartment and may become oxidized into progesterone-receptor active metabolites and may, therefore, alter gene transcription.

As a consequence of the wide range of possible interactions of steroids with neuronal transmitter systems, neuroactive steroids have potent neuropsychopharmacological properties, including sedative, hypnotic, anesthetic, anxiolytic, antipsychotic, and anticonvulsant effects. With respect to estrogens, the antagonistic effect of 17β-estradiol at the NMDA-receptor may also contribute to estrogens neuroprotective activities in certain disease states (Weaver et al., 1997). The stimulatory effect of 3a-reduced pregnane steroids or of progesterone at GABA receptors may potentially have sedative, hypnotic, anaesthetic, anxiolytic and other effects. For a detailed review of the direct behavioral effects of neuroactive steroids the reader is referred to recent reviews on this topics (Frye and Duncan, 1994; for review: Rupprecht, 1997; Baulieu, 1997; Rupprecht and Holsboer, 1999)

Of course, above mentioned experimental findings may have important implications for the treatment of neuronal disorders. One may think here for instance about the interaction of GABA-receptors with estrogen with respect to the treatment of epilepsia, where the $GABA_A$-receptor is currently the major pharmaceutical target. But not only pharmacological effects need to be considered. Further, the potential of neuroactive steroids to function as endogenous physiological modulators of neuronal function has to be taken into account. Rapid changes in the level of various neuroactive steroids including progesterone and 3α-reduced pregnane steroids may contribute to alterations in mood and the discomfort of women during pregnancy, the premenstrual syndrome and the post-partum depression (Wang et al., 1996). Neuroactive steroids play a major role in the maintenance of the body's overall physiological homeostasis and pathological changes in the bioavailability that leads to a disturbed homeostasis and may ultimately increase the risk for the development of psychiatric syndromes and disorders (for review: Rupprecht, 1997). Consequently, it can be argued that the drop in estrogen levels after menopause may immediately alter the above mentioned physiological interactions. It is well acknowledged that one major characteristic menopausal change is manifested in the state of mood. But do these changes directly contribute to the development of degenerative disorders? Or do these changes set the stage for an increased sensitivity? Due to the possible neuropsychopharmacological and endogenous effects of physiological levels of neuroactive steroids, interactions with psychopharmacological drugs need to be considered as well.

cholinergic, serotonergic, and catecholaminergic system. In addition estrogen may also act on glial cells. Estrogen induces genomic and non-genomic events in nerve cells. Interactions with the cellular membrane are common as well as the interaction of estrogen with intracellular signaling pathways, such as the MAP kinase signaling. Circulating sex hormones account for various sex differences in brain structure and function. Most importantly, gender differences are prominent also in the hippocampus, which may account for different cognition and memory when women are compared to men. A drop of the estrogen levels after menopause is, therefore a dramatic event not only for the sex organs but also for the brain.

Many activities in the brain are due to the classical genomic mode of estrogen action via the activation of intracellular ERs. But also numerous non-genomic and frequently very rapid effects of estrogens have been observed. These can be caused, for instance, by interaction with membrane-associated receptors, such as ligand-gated ion channels and with neurotransmitter receptors. Because of these particular activities estrogen is a member of the family of "neuroactive steroids".

Via "cross talk" with intracellular signal transduction pathways and transcription factors estrogen can indirectly modulate gene transcription and can regulate the transcription of genes that do not contain EREs.

Lastly, estradiol is a powerful antioxidant and oxidative stress is a very prominent hallmark of neurodegenerative diseases in nervous tissue. Interestingly, the brain is particular vulnerable to oxidations. The protection of nerve cells against oxidative stress and the modulation of the intracellular redox-homeostasis may protect cells and may again, indirectly, modulate gene transcription.

In summary, the neuromodulatory activities of the female sex hormone are numerous and some of these effects may also serve neuroprotective functions. Due to that obvious importance of estrogen also for brain structure and function, it is understandable and very likely that a loss of estrogen during the female menopause may have dramatic functional consequences.

7. Protection of the brain by estrogen

At various sites of the book so far links have been given to neurodegeneration and neuroprotection. In the following chapters two main questions are addressed. Firstly, does estrogen protect the brain? Secondly, how is the protection mediated and can we learn something for the development of future therapeutics? The main summary of the last chapters is that estrogen, the steroid, the sex hormone, has so many activities via so many mechanisms. By asking for protection through estrogen one refers to estrogen almost as to a certain drug rather than to a physiological modulator of the body's homeostasis. Can estrogen protect neurons against life-threatening challenges? Should estrogen be applied for prevention, for treatment of disease and if, when should this be done? Can estrogen be used just like an immunization against a viral infection to protect against disorders that are known to be associated with an estrogen-deficit?

Estrogen as drug for the brain?

Given the described complexity of potential estrogen actions in the brain one can expect no simple answer for the question whether one could use estrogen for prevention and therapy yes? or no? Here the main focus is the central nervous system as a target for estrogen and neurodegenerative disorders. With respect to neurodegeneration the focus is AD since it is the primary cause of dementia in the elderly. There is no effective treatment for this disease. The demand for a therapy of AD is huge since the average life expectancy constantly increases in the industrial countries.

Estrogen is circulating through the bloodstream in the female body. The blood and tissue levels of estrogen are changing depending on the menstrual cycle. Once the menstrual cycle ceases e.g. after the menopause the estrogen levels rapidly drop. Besides the age-dependent occurrence of the menopause the removal of the uterus (hysterectomy) or of the ovaries also leads to a stop of the menstrual cycle associated with decreasing estrogen levels. Given the manifold activities of estrogen in the body and in the brain, in particular, such a dramatic physiological change must have tremendous consequences.

So it is well known that after menopause not only the monthly menstrous cycle and, therefore, the capability of conception has ended but also other organs and tissue that are massively affected by high estrogen levels before menopause and may experience dramatic changes as well.

Menopause, and estrogen replacement therapy (ERT)

Before we discuss the substitution of the female body with the female sex hormone by an hormone replacement therapy (HRT) with estrogens (ERT), one has to understand why there is a need for replacement with a hormone in the first place. Of course, there are several non-physiological reasons why a body needs extra estrogen. The most frequent case is the surgical removal of the uterus or ovary ducts for instance necessary during cancer intervention. Here only the non-pathological and physiological causes for the drop in the bodie's estrogen levels are discussed. What happens during "menopause", the cessation of menstruation in women. To address this one has to go back in the female lifetime. The ovaries of a newborn girl contain approximately 400 000 primordial follicles in the ovaries. During a normal reproductive life span less than 400 follicles will ovulate and the excess follicles will degenerate via a process called astresia, a mechanism related to apoptosis (for review: Palumbo and Yeh, 1995). Apoptosis is the physiological way to remove excess or dysfunctional cells. Under certain pathological conditions the induction of apoptosis in the wrong place and at the wrong time can lead to the removal of functional cells. Apoptosis is frequently observed in neurodegenerative disorders, including AD (for review: Behl, 2000). In developed countries the menopause occurs at an age of 51 in average. Interestingly, while this average age has been rather stable throughout centuries, the age at onset of menses, the menarche, has continuously declined. This earlier onset of menarche may be due to an improved nutrition and the successful prevention of systemic diseases.

Women after menopause are called postmenopausal just to indicate this fact. But before the menstruation finally stops there is an up to five year period before and after this pause which is referred to as "climacterium". So by definition the term menopause describes the latest spontaneous menstrual period and the term climaterium means the time span of the transition from the productive (pre-menopausal) to the postreproductive (postmenopausal) period.

The primary deficit at menopause is ovarian and the reduced ovarian function is accompanied with a cessation of ovulation and, of course, a decline in estrogen (and androgen) production. Once the ovulation is stopped, the reproductive capability is ultimately lost. Therefore, at menopause the ovaries and finally the whole female body undergo dramatic changes. The most impressive endocrinological change is the rapid decline in estrogen production and in the estrogen levels. It is known that aged ovaries (before menopause) produce less estrogen and is also less reactive to FSH, which stimulates the estrogen production and ovulation. Moreover, during climacterium also the androgen levels decline and with it also the secretion of precursor molecules of estrogen such as dehydroepiandrosterone (DHEA). Therefore, as estrogen levels decrease the levels of these precursors, which may be converted to estrogen in peripheral tissues, also decrease (Longcope et al., 1986). Estradiol secretion in women changes from 40–80 μg/day to 0–20 μg/day, while the decrease in androgen secretion is less prominent (Longcope, 1986). In addition, the plasma levels of estradiol change while the levels of the stress hormone cortisol do not change significantly.

As already described in previous chapters, hormones are highly potent biological signals acting in the nanomolar range in the cells. Considering this and the fact that estradiol may interact with various pathways in cells, dependent and independent of estrogen receptors, it can be easily imagined that such a dramatic drop in the level of a hormone (in many cases down to zero) may have significant physiological and systemic consequences. Therefore, there is a whole range of overall gross changes in the female body during climacterium and menopause. The obvious unpleasant symptoms include:

- Vasomotor Hot Flushes,
- Urogenital Atrophy,
- Vaginal Dryness,
- Depressive Symptoms.

Hot flushes are the most prominent signs of menopause and are described as sudden and transient body sensitivity giving the impression of warmth to intense heat all over the body. Over 75% of menopausal women (in the U.S.) report such hot flushes. Although still somewhat under debate menopause has been linked to changes in mood, memory, and sexual functions (for review: Hammond, 1996; Pearlstein et al., 1997).

Despite the fact that these changes are indeed unpleasant, there are much more dramatic consequences slowly occurring and accumulating over time in postmenopausal women. Long-term estrogen deficit after menopause may lead to

- general atrophy of the skin,
- enhanced bone loss from the skeleton that may cause osteoporosis,
- changes in the vascular system leading to an increased incidence of coronary heart disease,
- major changes in the central nervous system.

In the USA more than 250 000 hip fractures in females are related to osteoporosis and it is certain that low bone mass is a distinct risk factor for fractures at various sites of the skeleton (Ross et al., 1991; Slemenda et al., 1993). It is now known that bone tissue is in a balance between degradation and repair and in osteoporosis the breakdown activity is increased. Osteoporosis is defined as a reduction in bone mass, leading to fractures after minimal trauma. Novel approaches for the treatment of osteoporosis including ERT and the molecular implications for the pathogenesis of osteoporosis are under intensive investigation (for review: Manolagas, 2000).

The other major issue is the increased incidence of cardiovascular disease after menopause and indeed, coronary heart diseases are the leading causes of death among women in the U.S. as well as in Europe. The protective effects of estrogen in the prevention of atherosclerosis (arteriosclerosis) and cardiovascular disorders is discussed below.

Concerning the impact of an estrogen deficit on cognitive skills there is an ongoing debate and many data accumulated indicating that ovarian factors, and here mainly estrogen, are of "prime importance in the normal maintenance of brain function, and the loss of ovarian steroids at menopause may play a major role in the cognitive decline and neurodegeneration that are associated with Alzheimer's Disease" (Simpkins et al., 1994). Although many studies employed rodents that have been ovaryectomized (to remove the source of estrogen) various human studies are now also available. Barbara Sherwin recently elegantly summarized and evaluated a collection of investigations published over a long period of time and, indeed, she reports that "estrogen can maintain some aspects of memory in females" (Carlson and Sherwin, 2000).

Due to an increased life expectancy in the developed countries the average women lives about one third or even more of her life in the time period after the ovaries stopped the production of estrogen, postmenopausally. Recent calculations estimate that about 20% of the entire population (in developed countries) is represented by women either in the transition to menopause (climacterium) or already after menopause. It is very difficult to assess how many of these women are undergoing HRT/ERT but a general feeling in the medical community is that the numbers are constantly increasing. A population-based survey in Scotland revealed that 19% of women aged 45–54 were using HRT (Porter et al., 1996). The causes for this increasing numbers of HRT users are very likely manifold. Obviously the whole life philosophy has changed in the last decades and the ever lasting youth and beauty is one of the ideal pictures of a person propagated by the mass media. Also, there is a more open discussion about health problems, beauty and changes in sexual behavior and sex physiology so that the "outing" that a women takes estrogen for hormone replacement is no longer an exciting news. Perhaps rather the fact that a women does not take hormones might be surprising. Why avoid talking about HRT when everyone talks freely about sexual orientation and the use of Viagra (e.g. Lee, 2000). Obviously, this discussion is very special and should not be continued here but nevertheless it is interesting to note a general change in public awareness concerning health problems. It is very interesting how quickly a basic researcher gets involved when working on scientific problems that are related to sex hormones. In working with the female sex hormone estrogen and its activities in nerve cells and the brain opens immediately a debate about the pros and cons of ERT. This is a challenging situation since every scientist – at least those whose work is financed by public and governmental money – should be able to explain the impact and importance of his or her work on the society, since it is the society that pays much of the research. The aging society demands many answers to important health problems, mainly also with respect to neurodegenerative disorders.

ERT for age-related degenerative diseases: general remarks

What is aging scientifically? Aging was defined by Denham Harman, one of the pioneer scientists in aging research, "as the accumulation of diverse changes that increase the risk of death" (Harman, 1998). Many clinicians assume that there is quite a difference between diseases seen in the elderly and the naturally occurring aging process. Therefore, dementia that occurs at the age of 60 or 70 is very likely due to AD but dementia at the age of 100 is referred to as "natural aging". We age because the mechanisms of maintenance and repair in cells and tissue fail to preserve the normal structure and function. There is a great body of literature on theories of aging and with respect to the postulated mechanisms of CNS aging various factors are suggested to play a role, such as:

- the instability of nuclear and mitochondrial genomes,
- neuroendocrine dysfunction,
- increased oxidative stress,
- changes in the calcium homeostasis,
- and inflammation.

Again novel state-of-the-art techniques of molecular biology enable researchers to address now a complex panel of genetic changes in the aging process including the most relevant genetic companions of aging. With so-called *Expression Profiling* employing DNA-arrays the mRNA expression pattern of various areas of the brain between young and old animals were compared. A gene expression profile characteristic for the occurrence of inflammation, oxidative stress and reduced neurotrophic support was found for the old brain. Interestingly, these changes observed during brain aging in mice display many parallels some of the pathogenetic markers and features of neurodegenerative disorders (Lee et al., 2000). These data challenge the above mentioned view that there is a clear separation between natural aging and age-related disorders. On the other hand this approach also identified genes which mediate processes typical for aging. This means that a collection of clear-cut genetic targets are available. The regulation of the transcription and activity of these genes may affect age-related changes. Specific pharmaceutical inhibitors or activators of the relevant genetic programs can be designed. It is very unlikely that its one gene that controls the network of modulators of the various age-related events. Therefore, with the detection of age-associated gene expression the next step is the so-called *Functional Genomics*, which means here to decipher the interactions of the gene products and their functional consequences. Expression profiling, DNA-arrays and Functional Genomics will be discussed at the end of the book.

Estrogen affects a variety of neuronal systems in the brain via different mechanisms. We are aware that we are far away from understanding the full picture of potential estrogen target genes. Via the interaction with intracellular signaling pathways, estrogen can affect not only genes that contain typical EREs but can also modulate the transcription of genes that contain other transcription factor-responsive elements. Therefore, it can be expected that in any case estrogen very massively alters the expression profile of a neuronal cell. First results performed on a pure cellular level *in vitro* have confirmed these expectations (Garnier et al., 1997). Before we go into detail into possible cellular, molecular, and biochemical mechanisms of estrogens potential neuroprotective effects, we should get an impression about the basic advantageous and protective effects of estrogen in the body and here of course mainly on brain function such as cognition. In a next step, the current knowledge on the effects of estrogens in neurodegenerative disorders are reviewed. Perhaps, in addition to estrogens direct effects on the CNS activities in the periphery may also support estrogen's potential effects in the brain. One example is blood flow and circulation since the brain is almost exclusively dependent on the oxidative phosphorylation and uses glucose as the main source for cellular energy. Therefore, I will start the general discussion of estrogen's beneficial effects in the body with a view into estrogen and the cardiovascular system focusing on arteriosclerosis. Moreover, both epidemiological and neuropathological studies have suggested an association between Alzheimer's disease and several vascular risk factors, such as hypertension, inheritance of the apolipoprotein E epsilon 4 allele, coronary heart disease, diabetes mellitus, ischaemic white matter lesions and generalized atherosclerosis (for review: Skoog, 2000).

Estrogen and human diseases: general beneficial effects of estrogen

Estrogen and arteriosclerosis

The incidence of cardiovascular disease differs between men and women, which is due to differences in risk factors and hormones, at least in part (Barrett-Connor, 1997). The main hormonal player in cardiovascular disease including arteriosclerosis is the female sex hormone. The role of estrogen in the prevention of arteriosclerosis is very convincing since the incidence of this disease is rather low in premenopausal women, increases in postmenopausal women, and, most interestingly, decreases again (down to premenopausal levels) in postmenopausal women under ERT replacement therapy (Stampfer et al., 1991; Grady et al., 1992; Barret-Connor, 1997).

Therefore, it can be concluded that estrogen, indeed, is a preventive factor against arteriosclerosis.

The target tissue of estrogens potential protective effects includes also the cardiovascular system (Fig. 1). Interestingly, arteriosclerosis is an age-associated disorder and epidemiological studies suggest that vascular risk factors may be involved in Alzheimer's Disease (AD) as well as dementia in general (for review: Kudo et al., 2000). But what is the cellular and molecular basis of estrogen's preventive activity in arteriosclerosis?

Smooth-muscle cells and endothelial cells, the basic constituents of blood vessels bind estrogen with high affinity and the expression of ERα in these cell types has been demonstrated (Karas et al., 1994; Losordo et al., 1994; Venov et al., 1996; Kim-Schulze et al., 1996; Caulin-Glaser et al., 1996). With more and more investigations to be finished in the near future also the expression of ERβ will be determined. So far it is known that ERβ is expressed in blood vessels of mice and rats. In addition, both ERs are present in myocardial cells, which may be also of great functional significance. On the basis of the available data emplyoing ERKO mice it can be concluded that each of the two ERs is sufficient to protect against vascular injury (Iafrati et al., 1997; Lindner et al., 1998). It is well accepted that estrogen has various general systemic effects on cardiovascular functions. Estrogen affects serum lipid concentrations, coagulation and fibrinolytic systems, antioxidant systems, and the generation of vasoactive molecules (e.g. nitric oxide, prostaglandins). Interestingly, all these estrogen target molecules and mechanisms can significantly affect the development of vascular disease (for review: Mendelson and Karas, 1999). With respect to the antioxidant effects of 17β-estradiol it is known that this phenolic compound prevents the oxidation of the low-density-lipoprotein (LDL) in plasma *in vitro* (Shwaery et al., 1997; Santanam et al., 1998; Moosmann and Behl, 1999; for review: Parthasarathy et al., 1992; Steinberg, 1997). In postmenopausal women estradiol also had a significant effect on the LDL oxidation. Both long-term and short-term administration of 17β-estradiol decreases the oxidation of LDL cholesterol in postmenopausal women (Sack et al., 1994).

The literature that relates to lipid and lipoprotein oxidation to atherosclerosis has expanded dramatically in recent years. The "oxidative modification hypothesis" of atherogenesis has been supported by basic studies of the chemistry and enzymology of LDL oxidation, through studies of the biological effects of oxidized LDL on cultured cells, and also *in vivo* and with clinical studies investigating the effects of antioxidants on atherosclerosis in animals and humans. Nevertheless, despite a lot of data providing evidence for the beneficial effects of antioxidants in models of arteriosclerosis, there are still numerous contributing factors necessary to be studied and understood before antioxidant therapy becomes an option for the systematic treatment of cardiovascular diseases (for review: Parthasarthy et al., 1999; Chisolm and Steinberg, 2000).

In a recent *in vitro* test tube study the preventive effect of various phenolic compounds with and without estrogenic/hormonal activity have been investigated in LDL-oxidation assays. We found that it is the mere antioxidant activity that depends on the basic phenolic structure that mediates this preventive effect with respect to the LDL-oxidation paradigm. Compounds such as trimethylphenol (TMP) having no estrogen receptor-binding or -activating potential blocked the Cu^{2+}-induced LDL-oxidation just as effective as 17β-estradiol and is therefore completely independent from the activation of ERs (Moosmann and Behl, 1999). TMP is a highly interesting compound since it is a small molecule and, therefore, carries some basic features for good penetration of the BBB.

Estrogen can cause short-term vasodilatation, which is a rapid effect that does not involve genomic changes. It is believed that this rapid vasodilatory activity can be mediated either by estrogen's effects on ion-channel function or on the generation of nitric oxide (for review: Mendelsohn and Karas, 1999). But since ERs are expressed in the blood vessel tissue certain genes are typical target genes for estrogen including the genes coding for central vasodilatory enzymes such as nitric oxide synthase (NOS) (Weiner et al., 1994; Binko and Majewski, 1998). An altered expression of the gene for the inducible form of nitric oxide synthase may directly change the availability of nitric oxide (NO) in the blood vessels, and NO is the central relaxation factor (for review: Ignarro, 1999). The synonym for NO is *endothelial relaxation factor* and its synthesis through NOS is not restricted to the endothelium but is also occurring in the brain. In the nervous system NO is acknowledged as an important signaling molecule. In addition to all these activities, estrogen has also beneficial effects on the response to vascular injury.

It has to be stressed here that the biological effects of estrogen in the cardiovascular system may occur dependent and independent of ER activation and may range from alterations in well-known vascular estrogen target genes to the structural antioxidant activity of 17β-estradiol. Again, it is the wide range of estrogen's activities that may effect the cardiovascular system at various levels. Clinically, there is abundant evidence from both prospective and retrospective studies that ERT reduces the primary risk of cardiovascular disease in postmenopausal women without previous disease history. The extent of risk reduction may reach 35 to 50 percent and is therefore of great significance (Grodstein et al., 1996; 1997). The results of the Women's Health Initiative Memory Study (WHIMS) also will give insights into a possible role of estrogen for the primary prevention of coronary heart disease. Another target tissue of estrogen is the bone and the WHIMS also gives insight into the possible prevention of osteoporosis by estrogen. The influence of estrogen on bone physiology is well described.

Estrogen and osteoporosis

It is a fact that during human aging the loss of height is caused by a progressive loss of bone mass. The decrease in bone mass in combination with disturbances of the architecture of the skeleton leads to increased bone fragility and increased frequency of fractures. This condition is known as osteoporosis (Riggs and Melton, 1986). While both men and women experience a loss of bone mass starting around the age of 40, this loss is much more rapid in women. During the first 5–10 years after menopause, this process is most dramatic due to the decrease in estrogen levels. In addition to this dramatic effect after the female menopause, it is well known that during body growth women do accumulate less skeletal mass than men. This is the reason for smaller and thinner bones in women compared to men. The overall result of this gender differences is that the incidence of bone fractures is 2 to 3 fold higher in women (Orwoll and Klein, 1995). Besides these natural preconditions and the changes occurring during the aging process a loss of bone mass can also be caused by other factors, including

- chronic glucocorticoid excess,
- hyperthyroidism,
- alcoholism,
- long-term immobilization,
- cigarette smoking.

Bone loss in women needs to be assigned to two distinct events, the bone loss during normal aging, which similarly effects men, and the loss that is due to menopause. One has to realize that the bone tissue in the adult skeleton is still a highly active tissue. It regenerates by temporary cellular structures that comprise two types of juxtaposed tissues, osteoclasts and osteoblasts, and periodically replace old bone with new. A large body of evidence during the last decade has shown that the rate of genesis of these two highly specialized cell types, as well as the prevalence of their apoptosis, is essential for the maintenance of bone homeostasis. There is a constant balance between the generation and death of cells as in various other regenerative tissues. The average life span of osteoclasts is approximately 2 weeks and that of osteoblasts is about 3 months. Both types of cells die by the physiological form of programmed cell death, apoptosis. It is believed that common metabolic bone disorders, such as osteoporosis, result largely from a disturbance of this fine-tuned equilibrium of birth and death of these cells. The basic regulatory mechanisms and their consequences for disease is summarized in a very recent overview (for review: Manologas, 2000). Estrogen and androgen deficiency increases the apoptosis in bone tissue and in various experimental approaches in mice ERT could reverse this effect (Tomkinson et al., 1997; Noble et al., 1997; for review: Noble and Reeve, 2000). Estrogen appeared to have an anti-apoptotic effect. This preventive effect was investigated *in vitro* and, indeed, 17β-estradiol inhibited apoptosis of osteoclasts and osteoblasts. The expression of functional ERs seemed to be the prerequisite of this activity. But in contrast to the classical genomic ER pathway it was found that estradiol activated the MAP kinase pathway very rapidly in minutes. Therefore, it is not quite clear where ERs take part in this anti-apoptotic effect. And the finding that also the inactive steroid analog 17α-estradiol has similar activities leaves this question unanswered. Based on the studies performed in various cellular systems it may be the cytoplasmatic interaction of ligand-activated ERs that control the apoptosis preventing effect of estrogen in bone cells. This activity may therefore be called non-genomic or indirectly genomic.

After menopause a disruption of the balance of bone formation and bone resorption occurs. Current therapies for the treatment of osteoporosis, ERT, SERMs, and bisphosphonates, are primarily based on blunting and preventing the resorption of bone cells. For the treatment of osteoporosis SERMs could also be an option. Indeed, some SERMs provide bone protection without the estrogenic side effects seen in ERT (for review: Lopez, 2000). Recently, the SERM raloxifene has been shown to increase bone mineral density and decrease biochemical markers of bone turnover in postmenopausal women but lack the unwanted stimulatory effects on the breast and uterus. To investigate how changes with raloxifene compare with estrogen, a randomized, double blind study employing raloxifene or conjugated equine estrogens, which are the preparation of choice for ERT, was performed. The results of this study suggest that the SERM raloxifen reduces bone turnover but increases bone density, although to a lesser extent than the conjugated equine estrogen. Nevertheless, despite the better effects of the equine estrogens, it has to be considered that the SERM does not stimulate proliferation of cells in breast and uterus. And this is of course an important issue when considering potential side effects of ERT (Prestwood et al., 2000).

Estrogen as a drug for the treatment and prevention of brain diseases?

Effects of estrogen on cognition

Steroid hormones in general, can affect brain function. Many studies focussed on the interaction of stress hormones such as glucocorticoids on the CNS (for review: Holsboer and Barden, 1995; Lupien et al., 1999). Glucocorticoids (cortisol in man, also named hydrocortisone) are secreted by the adrenal gland as response to various types of stress. The secretory activity of the adrenal gland is the final step in the chain of events along the "hypothalamus-pituitary-adrenal (HPA) axis".

> The stress reaction starts in the brain and via different physiological pathways and players ends in the adrenal gland with the glucocorticoid secretion.

An "overshooting" stress response resulting in a so-called HPA-hyperdrive as well as a down-regulation of the stress response system can cause diseases (for review: Holsboer and Barden, 1996; Steckler et al., 1999).

Philipp Landfield intensively analyzed the influence of adrenal hormone on the aging of the brain and on the structure and function of the hippocampus. There is a body of literature on that particular topic. Of course, stress reactions and the associated secretion of stress hormones of the body are essential for our survival but an overload of stress may be damaging. Robert Sapolsky's and Bruce McEwen's work also emphasizes the pivotal importance of stress for our living on one hand and on the other hand the detrimental effects of glucocorticoids in the brain, mainly in the hippocampus. The hippocampus expresses rather high levels of steroid receptors for glucocorticoids and human studies show cognitive impairment consistent with hippocampal dysfunction in depression, bipolar disorder, and Cushing's disease. Alterations in hippocampal function are also reported in people that receive exogenous corticosteroids. Interestingly, alterations of the HPA-axis are also seen in patients with Alzheimer's Disease. Further, the role of glucocorticoids in cognitive functions of the hippocampus and in hippocampal nerve cell survival is of immense importance for the pathogenesis of Alzheimer's Disease (for review: Landfield, 1988; Landfield and Eldridge, 1994; Sapolsky, 1992, 1996, 1999; McEwen, 2000).

The sex hormones estradiol and testosterone may modulate cognition and memory. The hippocampus and adjacent cortical areas play a major role in memory function (for review: Squire, 1992). To "measure" memory, or rather, to determine memory function one has to use certain standardized tests. The most commonly used tests of so-called explicit memory are recall and recognition. These parameters show a major impairment in animals and humans with decreased sex hormone levels as well as in aging (for review: Craik and Jennings, 1992). The wide-ranging effects of estrogen on various brain structures and neurotransmitter systems that are directly involved in learning and memory have been introduced previously (see page 67). But what are the effects of estrogen on cognitive function throughout the female life-span? Due to the increased life expectancy and the increased amount of the post-menopausal period this is a central question.

It has to be stressed here again that in women before menopause almost all estradiol in the circulation (approximately 95%) is secreted by the ovaries (for review: Baulieu and Kelly, 1992). After menopause the ovaries become atrophic and stop secreting estradiol. Due to the fact that estrogen replacement is of increasing importance in postmenopausal women, various studies on estrogen's direct effect on cognitive functions in postmenopausal women have been performed. Many reports demonstrated a clear beneficial activity of estrogen in women who underwent a surgical removal of their uterus and ovary (surgical menopause, e.g. through hysterectomy).

In summary of several studies, it becomes evident that estrogen supplementation is beneficial for the maintenance of verbal memory but had no significant effect on visual memory (for review: Sherwin, 2000).

It has to be underlined that various sex differences in cognitive functions exist. There are gender differences in brain structure and function. The differences in cognitive functions are not detectable by the general rating of the intelligence quotient (IQ). It is known that men perform better in spatial and quantitative abilities as well as in tasks dependent on motor strength. Women on the other hand, on average have better

- verbal abilities,
- perceptual speed and accuracy, and
- perform better in tasks of fine motor skills.

The influence of estrogen administration in postmenopausal women on cognitive skills have been studied very intensively in groups of postmenopausal women who used estrogen compared to non-users. It is reported that the explicit memory, especially the verbal recall, was improved in estrogen-treated women (Campbell and Whitehead, 1977; Hackman and Galbraight, 1976; Kampen and Sherwin, 1994; Phillips and Sherwin, 1992; Robinson et al., 1994; Sherwin, 1998; for review: Carlson and Sherwin, 1998; Jarvik, 1975). In addition, in these groups estrogen also enhanced verbal skills, reaction time, attention, and work performance (Caldwell and Watson, 1952; Campbell and Whitehead, 1997; Fedor-Freybergh, 1977; Kantor et al., 1978). Of course, in such population-based studies, which are very difficult and tedious to perform, not all studies are completely consistent and variability does occur. Some of these contrasting findings may be the result of differences in the experimental approaches and the study design, meaning the number and exact type of neuropsychological tests performed or more practically the mode of application of estrogen (oral versus transdermal) and the exact doses. However, in summary the analysis of the association between estrogen and cognitive functions in postmenopausal women support the idea that estrogen helps to maintain short- and long-term memory.

Evidence exists clearly demonstrating that these obvious differences in cognitive skills can be due to the different prenatal exposure to sex hormones. Such evidence includes studies of individuals exposed to abnormal levels of sex hormones in fetal life due to genetic disorders. In addition to the investigations focusing on the prenatal exposure to sex hormones studies have been conducted concentrating on variations in cognitive skills during the menstrual cycle phases. Interestingly, women performed better on tasks of spatial abilities during phases of low estrogen levels. These data indicate that estrogen may have a negative effect on spatial performance and skills. In contrast to disadvantageous effects of estrogen on spatial skills, in periods of the menstrual cycle with high levels of estrogens verbal articulatory skills were improved (for review: Drake et al., 2000; Carlson and Sherwin, 2000; Sherwin, 2000).

Taken together, fluctuations of estrogen levels due to the menstrual cycle change cognitive functions. Low estrogen levels promote spatial skills and high estrogen levels support verbal skills.

As convincing as these findings on the cognition-improving effects of estrogens are, other studies were not able to show an effect of estrogen on cognition in menopausal women (e.g. Polo-Kantola et al., 1998). It has to be mentioned here that these conflicting studies may also be a consequence of the quite different methodologies used in the single investigations. These obvious menstrual cycle-dependent differences in cognitive function in women are exciting and one may speculate about possible consequences for daily living. Moreover, it has to be considered that many women take certain sex hormone preparations for contraception. What does that mean for the cognitive performance? Although this is a very interesting question, here it will be focused on the time period of menopause and the time thereafter.

To directly answer the central question "Can estrogen keep you smart?", recently, Barbara Sherwin undertook a great effort and reviewed and analyzed the biological plausibility of and the clinical empirical evidence concerning a link between estrogen levels and memory in women. The objective of this critical analysis was the clinical evidence published from 1952 until 1989. On the basis of the results of over 40 references she concluded that "Estrogen specifically maintains verbal memory in women and may prevent or forestall the deterioration in short- and long-term memory that occurs with normal aging. There is also evidence that estrogen decreases the incidence of Alzheimer's Disease or retards its onset or both" (for review: Simpkins et al., 1994; Sherwin, 1999; Henderson, 2000). It is evident that most of the studies concern with age-associated memory changes and some of them, of course, also with the most serious form of memory loss, dementia of the Alzheimer type.

In Germany a randomized double blind study analyzed the effects of a two week transdermal treatment with estrogen on memory performance in 38 healthy elderly women. The results of this study support the idea that estradiol replacement can have specific effects on verbal memory in healthy postmenopausal women. Moreover, it is suggested that these effects can occur rapidly with transdermal application of estrogen and that there may be a dose response relationship between estradiol levels and memory enhancement. The specific design of this study enrolling women who had been menopausal for an average of 17 years before entering this investigation furthermore indicates that the brain maintains a sensitivity and a plasticity for estrogens even after a longer time with low estrogen levels associated with the menopause (Wolf et al., 1999). This indeed is a highly interesting observation since it means that menopausal alterations in cognitive abilities may be partially reversed even late after the onset of menopause. This suggests that it is not a "now or never" decision to apply estrogen replacement at the time of menopause.

Recently, a large study was finished that measured cognitive performance with a modified mini mental status examination (mMMSE) in 425 women at the beginning of the study and then 6 years later at an age of 65 years or older. To determine the biological activity of estrogens, instead of the total hormone concentration, the non-protein-bound (free) and only slightly bound (bioavailable) forms of estrogen were determined with RIAs (Yaffe et al., 2000).

The radioimmunoassay (RIA) is a technique that employs an antibody which specifically recognizes a particular hormone and, therefore, binds to that hormone. Hormone-bound antibodies can be detected by using radio-labeled second antibodies that specifically recognizes the first antibody. Blood samples and other tissue extracts can be investigated by RIAs and makes the detection of very low concentrations of a hormone possible.

Of course, for the correct evaluation of the data of this study an adjustment for exact age, current estrogen use etc. was done. And it was found that women with high serum concentrations of non-protein-bound and bioavailable estrogen, but not testosterone, were less likely to develop cognitive impairment compared with women having low estrogen concentrations. This recent and very exciting finding supports the hypothesis that higher concentrations of endogenous estrogen, indeed, prevent cognitive decline (Yaffe et al., 2000). This large study further supports other findings indicating cognition-enhancing effects of estrogen. Most of the studies mentioned so far were derived from groups of women that experience decreased levels of estrogen due to the changes occurring with menopause. Therefore, the age-associated cognitive impairments were in the center of the presentation and in summary we learned that estrogen supports various cognitive functions. But is there a role for estrogen also in neuropsychiatric disorders, such as AD, where the progressive loss of cognitive abilities is the major clinical hallmark?

Estrogen in neuropsychiatric disorders

Neurodegenerative disorders – Alzheimer's Disease

Memory as a cognitive function enables us to accomplish our everyday living. Our life is not possible without recalling experiences and people, learning basic facts and recognizing people, objects and situations. Loosing memory means loosing your personal history and your personal identity. As cited in the foreword of this book, Shakespeare let his King Lear give the impression of lost memory. Most prominent Alzheimer Patients of our days are former U.S. President Ronald Reagan or the well-known actress Rita Hayworth. The progressive loss of memory and cognitive functions is the major clinical characteristic of Alzheimer's Disease (AD). AD is the most prominent cause of senile dementia and is further characterized by deterioration, and, ultimately, by the death of the patient. Even almost 100 years after the German Psychiatrist Alois Alzheimer first described his observation of a neurodegenerative disease of the brain, which was later called "Alzheimer's Disease" (Alzheimer, 1907), the diagnosis "definite Alzheimer's Disease" still depends on a combination of evaluations, on clinical data including neuropsychological test batteries, the general clinical anamnesis, and on the final histopathologic confirmation when studying the brain of deceased AD patients (post mortem studies). With the development of novel and more sensitive brain imaging techniques the diagnosis "probable AD", which is mainly based on the clinical evaluation in psychological tests and in interviews of the patient and relatives, is constantly improved. The great dilemma of why there is no early diagnosis of AD is still due to the lack of a valid and reliable disease-marker. Despite enormous efforts no such marker is available to date. Considering that AD has a preclinical phase of up to decades and that the disease is slowly progressing before obvious clinical symptoms, a marker that would indicate a "beginning AD" would be of invaluable help for a possible prevention and later also for novel therapeutic approaches (for review: Terry et al., 1994).

AD accounts for more than 70% of all late-onset cases of dementia. In the USA alone AD claims more than 100 000 lives per year. Given the increased life expectancy in the industrialized nations a further increase in the total number of people suffering from AD is expected. This means full-time care for the patients in the late stages of the disease and this effort will be an enormous personal burden for the caregivers and a financial burden for the society. Therefore, markers that specifically detect this devastating disease and ways of prevention and therapy are desperately needed. The public awareness regarding AD is constantly increasing. Various people of public interest suffer from this disease and the literature addressing the problems of the disease not only for the patient but, moreover, also for the caregivers, supports this awareness. AD can affect everybody. It is a deadly disease and there is no cure! Research to solve the obvious problems is intense and many landmark findings have been made since the first descriptions of AD by Alois Alzheimer. But none of these findings has led to the actual development of an Alzheimer-drug that is effective in the full range of patients, so far. But what causes this disease? Are there preventive approaches? What is the link to estrogen?

What is the cause of AD?

When investigating post mortem brain tissue of AD patients the major histopathological hallmarks of this neurodegenerative disorder can be found, which are:

- synaptic alterations,
- nerve cell loss primarily in cerebral cortex, hippocampus, and amygdala,
- deposition of extracellular amorphous amyloid β protein (Aβ),
- intracellular precipitation of hyperphosphorylated tau-protein.

The extracellular Aβ-deposits build up the so-called senile plaques. Hyperphosphorylated tau protein form the neurofibrillary tangles in the nerve cells (for review: Masters and Beyreuther, 1998; Selkoe, 1999; Haass and Mandelkow, 1999). Despite world-wide efforts for many years the exact pathogenesis of AD is still obscure and current therapeutic approaches are derived from various working hypotheses trying to explain the pathogenesis of this disease.

Various AD-hypotheses

One early view is expressed in the **acetylcholine-deficiency-hypothesis**, which is based on the fact that AD is associated with a massive decrease of the neurotransmitter acetylcholine (ACh) (Whitehouse et al., 1982). ACh is the main mediator of cognitive functions and is the central neurotransmitter necessary for cognition and memory. The demonstration of the substantial abnormalities in the cholinergic system in the brain of AD patients initiated the discussion on a possible pharmaceutical ACh replacement strategy. Similar to the approach that is still used to treat Parkinson's Disease via dopamine replacement, which was introduced in 1968, the aim in AD was to stabilize the cholinergic function. Such a stabilization can be reached at various levels. Therapies have been developed that focus on the prevention of the degradation of this neurotransmitter in the synaptic cleft maintaining therefore constant ACh-levels. ACh is physiologically degraded by the enzyme ACh-esterase. Developed AD drugs block this ACh-esterase and are therefore called ACh-esterase-inhibitors (see Fig. 27).

Another hypothesis is the **arthritis-of-the-brain-hypothesis** of AD, which concentrates on the investigation of the role of inflammation during AD-pathogenesis (Rogers et al., 1992). The main support of this hypothesis is derived from histopathology. AD brain tissue clearly shows a massive accumulation of inflammatory components, such as complement (McGeer et al., 2000; for review: McGeer and McGeer, 1999). Treatment of AD patients with anti-inflammatory drugs have been proposed and various clinical trials using such drugs have been already performed, others are currently in progress. Two initial preliminary clinical studies were rather optimistic in suggesting that anti-inflammatory drugs can decrease the rate of cognitive decline in AD. Follow-up studies further suggested that non-steroidal anti-inflammatory drugs (NSAIDs) cause a delay in the onset or slow down the progression of AD. In the following time the effects of NSAIDs have been studied in more detail. The most prominent activity of NSAIDs is the inhibition of cyclooxygenases (COX) and lipooxygenases, which may, ultimately, lead to the inhibition of prostaglandin synthesis and also to a block of ROS formation. The enzyme COX-2 is induced in response to the exposure of neurons to toxic concentrations of glutamate, suggesting a possible role for COX inhibitors in neuroprotection by counteracting the events following a glutamate insult. As a consequence of the growing body of evidence supporting a major role of inflammation during AD pathogenesis various anti-inflammatory drugs are tested or are planned to be tested in clinical AD trials in the future. Potential therapies include several drugs used in the treatment of inflammatory diseases. These may also include glucocorticoids that are in use for the treatment of rheumatic diseases as well as anti-malaria drugs. Based on the knowledge of the endangering effects of glucocorticoids on neurons employing these stress hormones may not be the best choice. The question is of course whether the pathogenesis of these typical inflammatory disorders are comparable with the pathological events leading to AD. AD is a slowly progressing disease and above mentioned inflammatory reactions are rather acute. Therefore, the basic pathogenetic processes in inflammatory disease may be only of limited value for the investigation of AD-related pathomechanisms. Nevertheless, AD pathology shows an inflammatory component (for review: Breitner, 1996; McGeer et al., 2000.

In summary, the actual outcome and the results of the trials employing anti-inflammatory drugs including some retrospective investigations on the relation of the use of NSAIDs and the incidence of AD are not completely consistent and, therefore, still under discussion (for review: Veld et al., 1998; Cacabelos et al., 2000; Eikelenboom et al., 2000).

A recent clinical trial reported on the improvement in neuropsychological tests performed by patients using anti-inflammatory drugs compared to non-users but, however, no significant differences in the amount of inflammatory glia, plaques, or tangles in either diagnostic group was observed. Therefore, these authors concluded that the long-term anti-inflammatory medication in AD patients enhances cognitive performance but does not delay the progression of the pathological changes (Halliday et al., 2000).

In addition, the **energy-depletion-hypothesis** of AD is also suggested (for review: Hoyer, 1998). This hypothesis summarizes the importance of a decreased glucose metabolism and, therefore, a decreased energy supply in the brain for the development of age-associated AD. Indeed, normal aging of the mammalian brain is associated with various metabolic changes and, of course, also with the drop in sex hormone levels. Metabolic disturbances include variations in the neuronal insulin receptor, the desensitization of the neuronal insulin receptor by circulating cortisol and receptor dysfunction subsequent to changes in membrane structure and function. These changes are responsible for the aberrations in glucose/energy metabolism (for review: Hoyer, 1998). Furthermore, this hypothesis is driven by the fact that the majority of AD cases are late onset forms with unknown etiology.

Here, it needs to be stressed that, although, various genetic alterations and mutations located on different chromosomes have been causally linked to some familial forms of AD (for review: Selkoe, 1999) age is the only reliable risk factor for the non-genetic sporadic forms of AD. Indeed, approximately 85% of all AD cases are late-onset (after the age of 65) non-familial, sporadic forms (Evans et al., 1989).

Various risk factors are under discussion, including head trauma, lower educational level, nutrition, and most importantly, the age process. As mentioned, age is accompanied by massive alterations in the brain glucose/energy metabolism. It is hypothesized that these accumulating dysfunctions build the stage for further even more dramatic pathological changes including, ultimately, also changes at the molecular levels (for instance, Aβ-deposition and hyperphosphorylation of tau protein). Therefore, it is the increased vulnerability of the brain, induced by very basic alterations as the background of the disease and as the driving force of the sporadic, strictly age-related forms of AD. The disturbance of the glucose metabolism during the age-process is a well-known phenomenon and this may influence the onset and progression of AD (for review: Finch et al., 1997; Hoyer, 1998).

Since age is the primary risk factor for the majority of AD cases (late-onset, sporadic), in analogy to the free radical theory of aging, the **oxidative stress hypothesis** of AD has been proposed (for review: Ames et al., 1993; Beal, 1995; Behl, 1999). Aging, in general, is associated with degenerative processes, including the degeneration of cells and tissues which can result in diseases, such as cancer, cardiovascular failure and neurodegenerative disorders. The free radical theory of aging, as previously mentioned, suggests that oxidative damage due to accumulating oxidative stress is the major player in the degeneration of cells and tissues (Harman et al., 1998). The role of oxidative stress for the aging process is still intensively investigated. The great interest here, at least in part, is that perhaps through the stabilization of the antioxidant defense systems longevity can be reached. One such measure would be the use of a certain diet or supplements, such as nutritional antioxidants and antioxidant pills. Recently the importance of the oxidative metabolism has been nicely demonstrated again. For this study the natural antioxidant system of the Nematode worm *Caenorhabditis elegans*, a frequently used animal model in basic research, was augmented by small synthetic superoxide dismutase/catalase mimetics. This treatment

The fatal consequences include the flooding of the brain tissue with blood. The affected brain regions are destroyed or crippled. Ultimately, ischemic stroke leads to cell death in the brain, sometimes in large areas. Stroke can be defined as the damage to a group of nerve cells in the brain, which is frequently due to interrupted blood flow, caused by a blood clot or a burst blood vessel. Depending on the area of the brain that is damaged, a stroke can have various clinical consequences including coma, paralysis, speech problems and dementia.

The full pathogenetic picture of stroke and the pathways of cell death are only partially known and many experimental approaches still focus on finding relevant target mechanisms for the development of novel stroke drugs. What is known is that brain injury following transient or permanent focal cerebral ischaemia (stroke) develops from a complex series of pathophysiological events that evolve in time and space. Stroke occurs due to haemorrhage or occlusive injury and results in ischaemia and reperfusion injury. A variety of destructive mechanisms are involved including ROS generation (oxidative stress), calcium overload, cytotoxicity and apoptosis as well as the generation of inflammatory mediators. Ischemia can be transient or permanent (for review: Dirnagl et al., 1999). Unfortunately, many clinical stroke trials have so far not been successful. Interestingly, stroke also has a genetic component and some rare forms of ischaemic stroke can be caused by a number of monogenic disorders. Cerebral autosomal dominant arteriopathy with subcortical infarcts and leucoencephalopathy (CADASIL), due to mutations in the gene for the protein *Notch 3*, recently received a lot of attention as a cause of familial subcortical stroke. The gene for notch is known from the fruit fly *Drosophila melanogaster* and the normal function of Notch is required in ectodermal cells to prevent the cells from differentiating as neuroblasts. Notch mediates intracellular signaling and has been implicated in various physiological processes (for review: Weinmaster, 1997). In addition to stroke, notch-signaling has been linked also to the pathogenesis of AD. The genetics and phenotypes of monogenic stroke are covered in a recent review (for review: Hassan and Markus, 2000). However, the majority of cases of ischaemic stroke are as multifactorial in etiology as is AD (see page 109).

Stroke and estrogen

There are sex differences in the severity of brain damage induced by transient ischemia and more general also in the overall response to lesions and challenges, and, last but not least, in the response to chronic stress (Hall et al., 1991; for review: Kimura, 1992). Increasing evidence demonstrates striking sex differences in the pathophysiology of acute neurological injury and also in its outcome. Indeed, in various experimental models females appear less susceptible to postischaemic and posttraumatic brain injury. Even more important evidence suggests that this gender difference applies to humans also. It is hypothesized that the greater general neuroprotection in females is likely due to the effects of circulating estrogens and progestins.

Numerous clinical studies have demonstrated some beneficial effects of estrogen against cerebral stroke in postmenopausal women (Finucaneet et al., 1993; Paganini-Hill, 1995). In various animal models the exogenous administration of estrogen has been shown to clearly improve the outcome after cerebral ischemia and traumatic brain injury. Both indirect and direct effects of estrogen may account for its neuroprotective effect and all of these protective mechanisms together will very likely provide a multifactorial protection. While the indirect effects of estrogen may involve the modulation of serum lipid levels and the reduction of platelet aggregation, the direct effects may be due to the complex range of estrogen's activities and may involve both genomic and non-genomic pathways. Again, estrogen may directly promote nerve cell survival via the activation of neurotrophins and neurotrophin receptors or via estrogen's structure-based intrinsic antioxidant effects against oxidative stress (induced by glutamate or Aβ). Other potential mechanisms which have been shown mainly at the experimental level will be discussed later (see page 142). Progesterone, which is also included in some stroke studies and which is also under discussion as a neuroprotectant, at least in some model systems, may exert such activities through general membrane stabilizing effects. Stabilization of membranes may serve to reduce the damage caused by lipid peroxidation, the main destructive reaction occurring during oxidative stress in biological systems. In cerebral ischemia a complex cascade of metabolic events is initiated and several of those involve the generation of free radicals and reactive oxygen species. It is strongly believed that these ROS mediate much of the damage associated with transient brain ischemia (El Kossi et al., 2000), and in the region surrounding the infarcts (penumbra region) caused by permanent ischemia. Various antioxidants have been used in animal models of stroke with varying results but, ultimately, when tested in clinical trials the outcome was not really successful (for review: Dirnagl et al., 1999). Ebselen, 2-phenyl-1,2-benzisoselenazol-3(2H)-one (PZ 51, DR3305), mimics the activity of the antioxidant enzyme GSH (see Fig. 21) peroxidase which also reacts with peroxynitrite and can inhibit a range of enzymes capable of inducing oxidative stress such as lipoxygenases, NO synthases, NADPH oxidase, protein kinase C and H+/K+-ATPase. Ebselen is in a late stage of development for the treatment of stroke (for review: Parnham and Sies, 2000). In summary, there is quite some evidence for gender differences in the outcome of stroke and also for the protective effects of estrogens (for review: Roof and Hall, 2000; Hurn and Macrae, 2000).

Frequently, in animal models of ischemic stroke estrogen is given prior to the insult. For instance the brain of ovarectomized rats supplemented with estradiol were less damaged by focal ischemia compared to non supplemented animals (Dubal et al., 1998). Another animal study employing the stroke paradigm of middle cerebral artery occlusion (MCAO) in rats demonstrated that both acute and chronic 17β-estradiol treatments protect the brain in experimental stroke (Toung et al., 1998). Interestingly, in this latter study, male animals were used for the experiments. In fact, it was shown that the male sex hormone testosterone did not alter estradiol-mediated tissue rescue after MCAO. Both latter investigations applied estrogen *prior* to the ischemic insult. But of course, ideal anti-stroke compounds would be effective also when applying them late after the seizure. This opens the time window for a possible therapy in which a beneficial outcome can be expected. Recently, an animal study was designed and performed by James Simpkins and colleagues to determine whether estradiol treatment after ischemia exerts similar effects as when given prior to the insult. Rats were subjected to permanent MCAO and E2 does indeed exert neuroprotectives effects when administered *after* ischemia. The therapeutic window in this permanent focal cerebral ischemia model was approximately 3 hours (Yang et al., 2000). This is an exciting result.

An interesting survey of literature availabe from 1960 until 1999 concerning the relationship between ischemic stroke and oral contraceptives revealed that the risk of ischaemic stroke is increased in current oral contraceptive users, even when low-estrogen preparations were used. This may have also other reasons since oral contraception has been associated with increased risk for e.g. thrombosis. However, the authors expect a rather small absolute increase in stroke risk since the incidence is very low in this population (for review: Gillum et al., 2000).

Schizophrenia

Besides the protective activity of estrogen, numerous reports ranging from molecular investigations to clinical studies also demonstrate the potency of estrogens to modulate central brain functions implicated in schizophrenia and depression. And indeed, estrogen's interaction with a variety of extrahypothalamic brain regions is manifested in alterations of dopaminergic, cholinergic, GABAergic, glutamatergic and serotonergic neurotransmission (for review: Alves and McEwen, 1999). Alterations in these transmitter systems are found in schizophrenia and depression.

Schizophrenia is a very heterogenous psychiatric disorder which includes most major psychotic disorders and is characterized by disturbances in form and content of thought, such as delusions and hallucinations, mood, sense of self and relationship to the outside world, and behavior, such as bizarre, apparently purposeless and stereotyped activity or inactivity. In schizophrenia gender differences were recognized early at the beginning of the last century by the German psychiatrist Emil Kraepelin. In principal, men and women have about an equal risk for developing schizophrenia but they differ with respect to the clinical manifestation of the illness. However, a survey of all available epidemiologic studies performed since 1980 revealed that in most studies the incidence of this disorder is higher in men. Women show:

- later age of onset,
- better premorbid adjustment,
- higher premorbid intellectual abilities.

In addition, the general treatment outcome for schizophrenia is much better in women (for review: Häfner and Anderheiden, 1997). While in men, the onset of symptoms of schizophrenia is most frequently between the age of 18 and 25, for women the age of onset is between age 25 through the mid-thirties (for review: Marsh and Casper, 1998). Nevertheless, women experience a second peak of incidence at an age older than 45 while men only very rarely develop this disease at this older age. It may be speculated that this second peak of schizophrenia incidence in women is due to changes in climacterium and menopause. Along with the changes in mood and an increased incidence in depression, the menopausal changes may also have an impact on the development of Schizophrenia. To put it very simple: early in life (when the estrogen levels are high in young women) women are somewhat protected against schizophrenia, later in life (when the estrogen level drops due to menopause) the schizophrenia incidence in women peaks.

While there is agreement that the main factor of the observed gender differences in the development and clinical profile of schizophrenia is estrogen (Seeman, 1997), the exact anti-pathogenetic role of estrogen remains elusive and is under discussion (for review: Seeman, 1996). Again, we face the same problem. Due to the manifold modes of action of estrogen in the brain and the possible interactions with various pathways of neurotransmission the exact role of estrogen is very hard to determine. One main view in the "estrogen hypothesis of schizophrenia" is that the reason for the later onset of the disease compared to men is the high estrogen levels in females that delays the onset in predisposed women until after completion of puberty. Therefore, estrogen plays a role as a protective factor, perhaps as a *psychoprotective hormone*. In fact, based on various clinical studies an "estrogen protection hypothesis of schizophrenia" was introduced (Cohen et al., 1999; for review: Seeman, 1996).

This view is similar to the proposed role of estrogen in AD where it is acknowledged that the increased estrogen levels as reached by ERT decrease the incidence of AD and delays the onset of this neurodegenerative disorder (for review: Henderson, 2000). Further arguments for a "protective" role of estrogen in schizophrenia is the fact that (1) the disease risk or relapse is much lower during pregnancy, when the estrogen level is high, and (2) the disease risk is increased when estrogen levels are low as before menstruation, during the postpartum period, or after menopause. As one possible mechanisms for this estrogen effect the modulatory activity of estrogen on dopaminergic receptor affinity is proposed. Indeed, estrogen downregulates the affinity of the dopamine receptor, an activity which may also promote the better outcome in women. As recently reemphasized, the neurobiological mechanism which can explain the delay of onset of Schizophrenia in women until menopause is very likely due to a sensitivity decreasing effect of estrogen on central dopamine D2 receptors (for review: Konnecke et al., 2000).

Estrogen has well known effects on mood, mental state and memory by affecting both "classical" monoamine and neuropeptide transmitter systems in brain. Various studies have demonstrated that E2 stimulates a significant increase in dopamine (D2) receptors in the striatum, which may also have implications for PD. In addition, estrogen increases the density of 5-hydroxytryptamine(2A) (5-HT2A) binding sites in various other brain regions including anterior frontal, cingulate and primary olfactory cortex and the nucleus accumbens, regions which are all implicated in the control of mood, mental state, cognition, emotion and behavior. These findings may also form the biological basis for the efficacy of estrogen therapy or 5-HT uptake blockers such as fluoxetine in the treatment of depressive symptoms of the premenstrual syndrome (see below). Furthermore, it suggests that the sex differences in schizophrenia may also be due to an action of estrogen on 5-HT2A receptors. Moreover, estrogen has potent effects on the expression of neuropeptides, such as arginine vasopressin. These manifold interactions of estrogen with various brain functions that are directly or indirectly related to schizophrenia may explain some of the gender differences seen in this disorder, as in others.

Depression

Depression is a generic term for a wide range of conditions and can range from normal feelings of the blues through dysthymia to major depression. Major depression is defined as a clinical syndrome that includes a persistent sad mood or loss of interest in activities that persists for some weeks. Further characteristic changes in behavior include altered eating habits, insomnia, early morning wakening, lack of interest, depressed mood, fatigue and suicidal thoughts. Women suffer from major depression two to three times more frequent than men. The reasons for the increased incidence in women are manifold. Genetic as well as biological and psychosocial factors may account for the increased incidence of depression in women. There are three main events in women's life related to changes in the hormonal equilibrium that carry an increased risk of depression (for review: Avis et al., 1994; Studd, 1997):

- premenstrual phase,
- postpartum stage,
- climaterium/menopause.

Estrogen supplementation in physiological doses can counteract premenstrual syndrome, the depression after birth, and depressive syndromes associated with the climaterium. Klaiber and colleagues demonstrated that ERT also has beneficial effects in major depression and significantly improved the mood of the patients (Klaiber et al., 1979). The outcome of ERT in climacteric depression was not really consistent and further clinical studies are necessary to analyze possible benefits of estrogens in depression, in general.

The pathogenesis of depression is not fully clear. Due to the great variety of disease conditions and also the multifactorial origin of this disease, there is – similar to other neuropsychiatric disorders, such as AD – no unifying hypothesis of its etiology. Besides a variety of neurotransmitter systems including norepinephrine, acetylcholine, and serotonin (5-HT), hormonal systems have been implicated in depression. Neuroendocrine changes in the thyroid hormone system as well as in the HPA-system play a central role in the pathogenesis of depression (for review: Holsboer and Barden, 1995). Moreover, the so-called catecholamine hypothesis, the adrenergic-cholinergic hypothesis and the 5-HT hypothesis are all supported by experimental and clinical data (for review: Cyr et al., 2000). With respect to the latter the decreased activity of 5-HT in depression has received increasing attention, which was due, at least in part, to the clinical success of the antidepressant fluoxetine, a 5-HT reuptake inhibitor. The neurotransmitter 5-HT is usually removed from the synaptic cleft, the site of neurotransmission, by reuptake mechanisms. Transporter proteins (5-HT transporter) are known and the inhibition of this reuptake by specific inhibitors such as fluoxetine increases the time that 5-HT is present in the synaptic cleft. Therefore, it is believed that decreasing levels of 5-HT can be compensated by prolongation of the lifetime of 5-HT in the synaptic cleft. A similar mechanism is the basis for the ACh-esterase inhibitors, which are used in AD therapy to extend the neurotransmission of ACh, the neurotransmitter whose levels decreases in AD. Again, it is the similar therapeutic approach as is proposed for the treatment of AD and PD, that is the supplementation of the CNS with a lacking neurotransmitter. We know today that all of these neuropsychiatric disorders are much more complex. The decrease in the level of certain neurotransmitters may reflect one single symptom rather than the initial cause of the disease. The role of 5-HT – or better – of the decreased levels of 5-HT in depression is further supported by other observations. For instance, a drop of 5-HT levels in the brain can induce depression in recovered depressed people and certain agonists of 5-HT that mimic 5-HT-induced neurotransmission can improve mood (Ananad et al., 1994). Possible interactions of estrogen with the serotonergic neurotransmission have been introduced before (see page 72). But does estrogen have an impact on the etiology or the progression of depression?

Women before the menopause show a significant dependence of the activity of the serotonergic system, e.g. the uptake of 5-HT, with the changing levels of ovarian hormones during the menstrual cycle (Halbreich et al., 1995). In post-menopausal women the level of 5-HT in the blood decreases just like the concentration of estrogen, and ERT can restore these levels. Recently, a large clinical study was finished that aimed to determine whether ERT in postmenopausal women is associated with fewer depressive symptoms. A so-called cross-sectional study with over 6000 women at the age 71 or over was performed. Based on the results of this study it can be concluded that "current use of unopposed estrogen was associated with a decreased risk of depressive symptoms in older women" (Whooley et al., 2000; for review: Marsh and Casper, 1998).

In various animal studies the influence of estrogen on players of the serotonergic system was demonstrated (for review: Rubinow et al., 1998). It can be concluded that the serotonergic system is modified by estrogen which may form the basis for potential antidepressant effects. Recently, an interesting link between estrogen, neurotrophins, and antidepressant activity has been proposed. Brain-derived neurotrophic factor (BDNF) may have antidepressant activities as found in animal models of depression, considering all limitations in trying to model such a complex disease as depression (Siuciak et al., 1997). Several antidepressant treatments increase the expression of BDNF and its receptor (trkB receptor) in the cortex and in the hippocampus of the brain. BDNF may mediate antidepressant effects by acting as a neuroprotectant that promotes survival and stabilizes the activity of 5-HT neurons. Like antidepressants estrogen modulates the expression of neurotrophin receptors (Sohrabji et al., 1994). Interestingly, the gene coding for BDNF contains a DNA sequence that is very similar to the ERE and estrogen increases its expression. From the fact that (1) estrogen may have anti-depressant-like effects, (2) certain anti-depressants increase BDNF activity, (3) BDNF gene contains a putative ERE, and (4) estrogen increases BDNF expression, a novel network of possible protective "cross-talks" can be envisaged. Although still speculative at the moment such a mode of action may pave the way for novel therapeutic approaches for the treatment of such a devastating and frequently deadly disorder like depression. Another possibility could be that BDNF does not act as a neurotrophin but rather modulates other mechanisms that are directly or indirectly related to the activity of neurotransmitter systems that may take part in the etiology of depression.

It is highly interesting to realize that different pathological conditions have a similar relationship to estrogen and to alterations in the estrogen levels. One has to remember that the neurodegenerative disorder AD is a disease of ACh-deficiency and estrogen has potent effects on the cholinergic system and its neurotransmission. Depression, where dysfunction of transmitter systems rather than nerve cell death is the key feature, can also be seen as a disease of decreased neurotransmission for instance through the serotonergic system, which is affected by estrogen. As for AD and schizophrenia, an estrogen-protection-hypothesis may also be proposed for depression.

After reviewing the literature with results on epidemiological and clinical studies aimed to investigate the effects of ERT on various neuropsychiatric (and non-neuronal) diseases, one has to stress that in all cases more studies with better controls and larger sample size are needed to get a much clearer picture. Nevertheless, it is safe to say that ERT is associated with various beneficial effects in humans including:

- a significant reduction in overall mortality,
- a reduction in the risk to develop cardiovascular disease or osteoporosis,
- increased perfomance in neuropsychological tests of memory and cognition,
- a reduction in the incidence of AD,
- protective effects in stroke models,
- a decreased risk of depressive symptom

(Hunt et al., 1990; Stampfer et al., 1991; Grady et al., 1992; Lafferty and Fiske, 1994; Henderson, 2000; Whooley et al., 2000). In summary, various clinical trials employing ERT for AD prevention demonstrate an improvement of cognitive functions in AD patients. In contrast to the studies on AD prevention, more recent AD treatment studies demonstrated no estrogen effect in various measures in people that are already affected (Henderson et al., 2000; Mulnard et al., 2000). With respect to AD, despite the preventive potential of estrogen, for disease treatment the female sex hormone may have too little effect too late (Marder and Sano, 2000). Therefore, estrogen can play a role for AD prevention but possibly not for AD therapy.

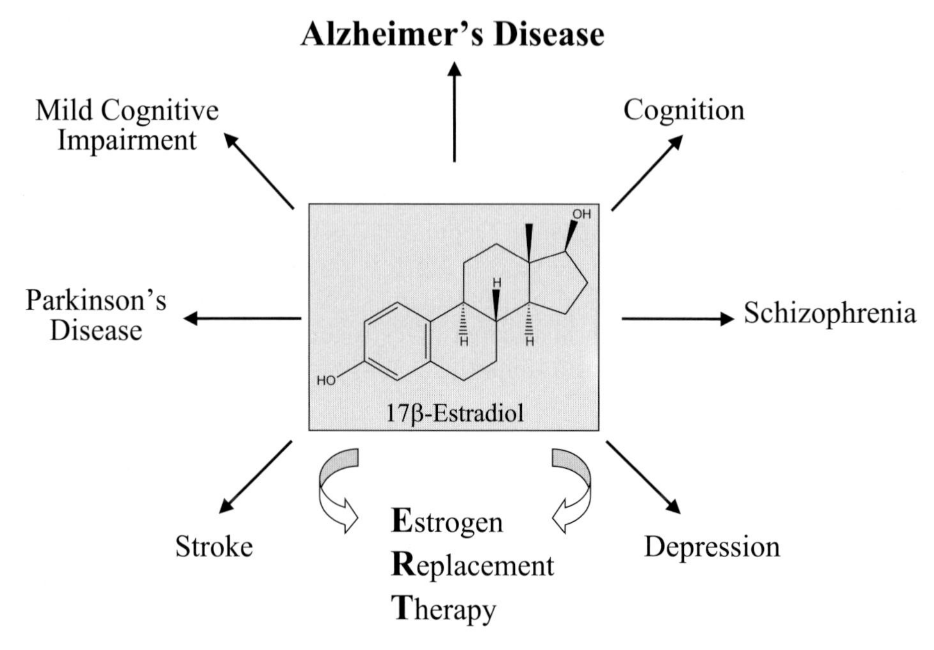

Fig. 38. ERT, brain function and neuropsychiatric disorders.
Estrogen interacts with various brain functions relevant for neuropsychiatric disorders. Estrogen's beneficial effects on cognition may play a central role in its preventive activities in AD

To replace or not to replace? ERT and breast cancer risk

Three different natural estrogens are used for ERT. Among the most commonly used is 17β-Estradiol in various formulations, which is also used in most *in vitro* and *in vivo* experiments. Estradiol can be administered orally, as a skin patch, as a gel preparation, or in form of a pellet, which may be implanted subcutanously. Very frequently, so called equine estrogens, which comprise a complex mixture of conjugated estrogens, or estradiol valerate are used. Both these compounds are taken orally.

Complience with any pharmaceutical intervention and therapy is a very complex issue and with respect to ERT currently the greatest concern for patients is that estrogen may increase the risk of breast cancer. There have been almost as many studies that show that ERT *increases the risk* of breast cancer as clinical studies that report *no significant effect* (for review: Beral et al., 1997). The results of further studies needs to be awaited. Nevertheless, I would like to underline that – and this is, of course, a personal view which evolved during the review of the literature and on the basis of many scientific discussions – the overall benefits of ERT outweigh its possible risks, especially when ERT is performed medium-term (up to five years). However, more specific drugs need to be designed that mimic the activities of estrogen in the desired target tissue and have no effects or no opposing effects in the breast tissue. Such new drugs are indeed on the horizon, some of them are already in clinical use, and are for instance SERMs. As outlined before SERMs mimic some of the activities of estrogen in the bone and cardiovascular system with respect to the prevention of osteoporosis and arteriosclerosis but act as ER antagonists in breast tissue (for review: McDonnell, 1999, 2000; Osborne and Fuqua, 2000). Next, some of the pros and cons of ERT with respect to breast cancer risk are presented.

Although the effects of estrogen on various neuronal (and non-neuronal) diseases are very complex and many interactions of estrogen with various mechanisms have to be observed, under certain conditions beneficial effects of ERT are well acknowledged. The decision to apply ERT is definitively not easy since the benefit and disadvantage need to be carefully considered. The major concern of women who face this decision is that estrogen may promote or even initiate human breast cancer. It would be a topic for one book alone to discuss these concerns and to present the experimental and clinical evidence. This has been done in some recent reviews which are cited within this chapter. Here, a short detour is made and some thoughts are given that are meant as a basis for further extensive discussion rather than as a full review. What is the biological basis for the possible increased risk for breast cancer caused by estrogens?

Most importantly it is well known that estrogens can act as mitogens for certain breast cancer cells, they promote the survival and the proliferation of these tumor cells. Consequently, this proliferative activity of estrogen and ERs have been investigated in many breast cancer models and have been implicated in the etiology of breast cancer (for review: Pike et al., 1983; Verheul et al., 2000). On the other hand the presence of ERα in the breast tumor tissue has been associated with a more favorable prognostic outcome in breast cancer patients (for review: McGuire, 1978; Verheul et al., 2000). Indeed, the proliferative effect of estrogens on breast cancer cells has been well studied and acknowledged for a long time (Knazek et al., 1977). Further, some experimental studies support the view that the expression of ERα is correlated with a reduced probability of forming metastasis (Thompson et al., 1992; Rochefort et al., 1998). The latter view is supported by recent findings suggesting that ERα protects against cancer invasion in its unliganded (no E2 binding to the ER) form and also after hormone binding although via a different mechanism (Platet et al., 2000). It may be argued that if the proliferation of a breast cancer is actually dependent on the activation of ERs, antagonizing these receptors by anti-estrogens prevent further tumor growth. One of the most widely and frequently used and effective agents in the treatment of breast cancer is the anti-estrogen, tamoxifen. Tamoxifen's efficacy is based on the specific biology of breast cancer with respect to ER expression. In fact, approximately two thirds of breast tumors express ERα. The proliferation of many of these tumors is indeed dependent on estrogen for growth and, therefore, these tumors respond to treatment with anti-estrogens. Unfortunately, the ER-negative remaining third of breast cancers generally do not respond to anti-estrogens (for review: Nass and Davidson, 1999).

In 1997 the "Collaborative Group on Hormonal Factors in Breast Cancer" has gathered and reevaluated about 90% of the epidemiological evidence that is available worldwide on the relationship between risk of breast cancer and use of hormone replacement therapy (HRT). It was concluded that the risk of having breast cancer diagnosed is increased in women using HRT and increases with increasing duration of use but that this effect is reduced after cessation of use of HRT and has almost completely disappeared after about 5 years (Beral et al., 1997). Indeed, when looking through the literature a major factor appears to be the duration of ERT. Many investigators underline an increased breast cancer risk in women which are long-term users of ERT and stress that this particular adverse effect overwhelms the other well-known beneficial effects of ERT.

In several recent reviews epidemiological and biological data on ERT and the risk of breast cancer risk have been discussed. Many conflicting experimental and clinical data exist and it is not possible to come to a consistent conclusion. Various aspects need to be considered in this discussion. One is the fact that the majority of epidemiological data are derived from studies in which the patients received high doses of estradiol (usually given as conjugated estrogens, estradiol conjugated to another group at C-17; see page 18). Another point is that, interestingly, in overweight women ERT does not increase the breast cancer risk, which could be a result of the fact that oral estrogens can reverse some characteristics of obesity. Obesity in itself can potentially increase breast cancer risk. Further, the use of androgenic compounds in ERT needs a special consideration, too. Also, one major point appears to be the right dose for ERT and it is argued that even the long-term use of low doses of oral estrogens (together with the addition of a non-androgenic progestin) could be associated with a limited increase in the breast cancer risk (for review: Campagnoli et al., 1999; Hammond et al., 2000). Due to the importance of this issue various surveys and evaluations of some of the available data were reviewed. Interestingly, another recent summary concluded that the preclinical and clinical data indicate that higher endogenous estrogen levels in the body that occurs for instance during pregnancy, or as reached through oral contraceptives and in ERT may be associated with an increased probability of breast cancer diagnosis. But on the other hand the data do not show that estrogens have negative effects on the course of breast cancer.

Taken together, the discussion of the pros and cons of ERT and the question of whether the beneficial effects of ERT outweigh potential health risks is an ongoing debate. It is a fact that large-scale randomized comparative trials of ERT in healthy women and breast cancer survivors are now ongoing but it will be some years before useful clinical information becomes available. Until then, one is advised to consider the limitations and potential biases that may occur as a result of the design of certain observational studies for the interpretation of such studies (for review: Marsden, 2000). As with many other pharmaceutical interventions, drugs have specific effects but also non-specific, side-effects. An ideal ERT with the goal of disease prevention would induce a potent stimulation of ERs, which – together with the other potent protective effects of estrogen – leads to targeted neuroprotection for instance in brain areas affected by neurodegenerative disorders (e.g. hippocampus) without affecting other areas expressing ERs in the body.

In summary, it can be concluded that there is no strong evidence to support the view that after the onset of breast cancer, estrogens worsen prognosis, promote or accelerate the course of the disease, or negatively interfere with the therapeutic management of breast cancer. Again, the outcome of several ongoing prospective, randomised clinical trials with different HRT regimens in healthy women or women who survived breast cancer needs to be awaited to allow better conclusions about risks and benefits of HRT (Verheul et al., 2000).

When considering the use of ERT it is indispensable to carefully look at each single woman who can safely benefit from hormone replacement. Some basic clinical factors and markers that need to be considered include obesity, bone density, mammographic density, circulating levels of androgen and estrogen, alcohol consumption, benign breast disease, and possible genetic factors (for review: Chiechi et al., 2000).

With respect to genetic predisposition factors, the identification of genetic mutations that are believed to be directly responsible for the breast cancer is one great advance for the dissection of possible molecular pathways of breast cancer development and, therefore, for the understanding of this disease. Approximately 5% of all breast cancer cases are thought to be caused genetically, the majority very likely by mutations in the BRCA1 and BRCA2 gene (for review: Warmuth et al., 1997). Interestingly, with respect to the 5% inherited disease cases there are similarities to the familial cases of AD, where about the same number of all cases are familial forms. Like in AD, the study of these mutations and the pathophysiological consequences will possibly reveal pathogenetic mechanisms which may apply also for the non-genetic cases. However, results of a population-based study weakened the support of the pathogenic role of BRCA2 somewhat by the finding that the mean risk of breast cancer in carriers of a mutation in the gene for BRCA2 is, indeed, lower than previously suggested (Thorlacius et al., 1998). The interaction of various genetic polymorphisms and the breast cancer risk has been summarized recently (for review: Dunning et al., 1999).

A successful treatment of those breast cancer forms, which depend on ER stimulation through estrogen, would specifically block tumor growth by antagonizing the ER. Anti-estrogens, such as tamoxifen, exist and are widely used for the treatment of breast cancer with some success. It is reported that tamoxifen induces a shrinkage of the tumor in advanced breast cancer, reduces the risk of relapse in women treated for invasive breast cancer, and prevents breast cancer in high-risk women. But tamoxifen has also interesting "side-effects" such as a beneficial effect in the prevention of postmenopausal osteoporosis (for review: Dhingra, 1999; Verheul et al., 2000; Breckwoldt and Karck, 2000; Brown and Lippman, 2000). These and other findings have led to the development of the concept of the design of an "ideal" SERM. As learned before SERMs are drugs with desired, tissue selective, estrogen-agonist or -antagonist effects. One example is raloxifen, which is known for its desirable mixed agonist/antagonist effects as mediated by tamoxifen but that lacks tamoxifen's uterine side-effects. Raloxifen was introduced into the clinic in 1998 (Delmas et al., 1997). In addition to raloxifen, certain SERMs are already investigated clinically (for review: Plouffe, 2000). The future design of an ideal SERM or a "designer estrogen" for the brain should transduce the tissue-specific neuronal actions and avoid, of course, untoward side effects for other tissues, such as the breast. The identification and evaluation of the mechanisms and genes that mediate neuroprotection and that are responsible for the beneficial effects in neuronal disorders are just in the process to be uncovered and it may take some time until this knowledge can be applied for a rationale drug design targeting these particular pathways. In general, drugs with an estrogen-like activity in the brain may have a great therapeutic potential either by modulating brain functions, such as neurotransmission, or through neuroprotective activity.

Menopause is an age-associated natural process that is mainly marked by the cease of ovarian function in women resulting in a rapid decline in estrogen levels. Due to the wide range of activities estrogen has throughout the body and also in particular on brain structure and brain function, estrogen deprivation is believed to, consequently, lead to various pathophysiological changes and disease conditions including osteoporosis, tooth loss, cardiovascular disease, stroke, age-related macular degeneration, colon cancer, diabetes mellitus, mood changes, higher incidence of schizophrenia, decreased cognitive functions, PD, and AD. Among these disorders, cardiovascular disease (coronary artery disease) is the most frequent cause of death in women age 50 years and older. Convincing clinical data suggest that women under post-menopausal estrogen replacement experience up to 50% reduced cardiovascular mortality. The issue of an increased breast cancer risk under ERT, the major concern of women who are facing ERT, is still under discussion. However, evidence suggest various beneficial effects of ERT on the CNS, including various neurotransmitter systems (e.g. ACh, 5-HT, dopamine) in the brain, which are implicated in learning and memory and AD (ACh), mood disorders and depression (5-HT), and movement disorders, e.g. PD (dopamine) (for review: Hammond et al., 2000). Finally, many beneficial effects of ERT have been repeatedly and convincingly described. However, the decision to use ERT after menopause is an individual one and has to be made considering various factors, such as familial predisposition to cancer or when the individual belongs to the group with an increased risk of breast cancer. The clinical trials showing beneficial estrogen effects in various diseases are numerous and many researchers believe that the benefits of ERT outweigh the potential risks by far.

8. Nerve cell protection by estrogen: molecular mechanisms

In recent years many experimental studies have focussed on estrogen's ability to prevent nerve cell death, which is the hallmark of progressive neurodegenerative disorders. These preclinical data strongly support the clinical data showing beneficial effects of estrogen for instance in AD prevention. The exact molecular and the cellular mechanisms of neuroprotection by estrogens need to be understood in order to optimize the protective effect of estrogen or of estrogen-derivatives. The answer to the question what are the mechanisms, what are the genes that mediate estrogen's neuroprotective effect against pathogenetic processes, will identify novel targets of pharmaceutical intervention. Next the currently discussed mechanisms that may mediate estrogen's direct protective effect in neuronal systems are presented and discussed.

Life is difficult, at the cellular and molecular level, too

As pointed out throughout this book numerous observations demonstrate that estrogen is a neuroactive compound and more than just a sex hormone. Estrogen influences brain structure and brain function. This is unequivocally shown by investigations *in vivo* and *in vitro*.

- Biochemistry demonstrates the structure of the molecule and the intrinsic structural potential of estradiol.
- Molecular biology defines the DNA that codes for estrogen receptors and their functional protein domains and describes the interactions of estrogen with its receptors.
- Cell biology employing cultured nerve cells show that estrogen has powerful neuroactivities at the cellular levels, ranging from the modulation of basic electrophysiological features to the activity to shield and protect the cells against oxidative damage.
- Physiology entertaining rodent and other animal models demonstrate the potency of estrogen in the body and its effect on various physiological measures ranging from sexual behavior to neuroendocrine functions.
- Clinical trials and many observations in humans underline the central

importance of estrogen for the physiology of the human body. Novel methods of brain imaging show the gross differences between the male and the female brain. Age-related diseases in postmenopausal women are studied in clinical investigations and the interaction of estrogen levels with disease states is analyzed. Various clinical trials are ongoing investigating the beneficial effects of ERT to prevent and treat age-associated degenerative disorders in women (e.g. osteoporosis, arteriosclerosis, AD). The loss of estrogen function that occurs naturally after menopause or by surgical removal of the ovaries before menopause clearly demonstrates the general impact of estrogen on the entire body structure, function and behavior.

As discussed before the female sex hormone is implicated in a wide range of pathological conditions, ranging from osteoporosis to neuropsychiatric disorders such as schizophrenia, depression, stroke, PD, and AD. Epidemiological data from many laboratories strongly support the view that estrogens provide neuroprotection of the CNS, which consequently may have implications for the etiology of neurodegenerative disorders. As pointed out the modes of molecular and cellular action of estrogen are numerous and basically all have to be taken into account when trying to study possible mechanisms of neuroprotection. Numerous genomic or non-genomic mechanisms of actions of estrogens in the brain have been documented implicating classical nuclear estrogen receptors as well as possible estrogen membrane receptors, antioxidant activity of the steroid 17β-estradiol, their effect on fluidity of the cellular membrane as well as on anti-apoptotic proteins and growth factors. More recently the development of molecules that specifically modulate ER function, the SERMs, further add to the spectrum of activities of the ERs. The design of SERMs that have the same beneficial effect as estrogen for instance in skeleton, cardiovascular systems and brain but act as antagonists in breast and uterus is one central area of hormone research of the future. Compounds could be developed that exclusively act in the CNS and protect the brain against age-related neurodegeneration. Before being able to design such structures one has to understand the protective mechanisms of estrogen in nervous tissue. The preventive activities of estrogen in AD and the basic cognition-enhancing effects are well known. However what is the biological, the cellular, the molecular basis of these estrogen actions? The protective activities need to be deciphered step-by-step for future drug design. We need to understand under which circumstances ERT or replacement with an estrogen-analogue is beneficial for the brain and when it is contra-indicated. Again based on the fact (1) that estrogen has activities in various target tissues throughout the body including the brain and (2) that many brain regions are targets for the action of estrogen including the hypothalamus but also various extra-hypothalamic sides, a complex range of neuroactivities has to be taken into account to explain estrogen's effect on memory, cognition, mood and other functions.

Very strong is the *in vitro* and *in vivo* experimental evidence for a nerve cell survival promoting activity of estrogen. Neurodegeneration means neuronal dysfunction and loss of nerve cells. AD is a neurodegenerative disorder with a complex etiology. Estrogen prevents the death of nerve cells in paradigms that are believed to be central to the AD pathogenesis. A detailed dissection of the mechanisms of nerve cell death on one hand and of the interaction of cell death with estrogen and ERs on the other hand is the basis for the approach to design "super-estrogens" or "super-SERMs" with optimized protective activities. The following chapters demonstrate the experimental evidence of estrogen's neuroprotective activity. Throughout, the relevance of the described activities for AD, the devastating non-treatable disease of the brain, is emphasized. Since various pathologies in the CNS are the consequence of similar pathogenetic pathways, many of the basic mechanisms defined for *in vitro* paradigms of AD may also apply to other neurodegenerative disorders.

Protective effect of estrogen in cultured neuronal cells

Mechanisms of nerve cell death

The basic mechanisms of cell death are complex and when looking at post-mortem brain tissue, for instance of AD patients, only the endpoint, the tombstones, of a long chain of pathological events leading to nerve cell death can be seen. It is not possible to study nerve cell death in the living human brain at the molecular level. For a detailed analysis of nerve cell death standardized model systems need to be used. A simple approach to prevent AD (and other neurodegenerative disorders) would be the prevention of degeneration, to stop nerve cell death. To do so, of course, detailed knowledge on the individual steps of cell death need to be clarified. What are the molecular changes that, ultimately, tell the cell to survive or to die?

Post-mortem studies revealed that both synaptic dysfunction and loss of nerve cells are among the most prominent histopathological features of AD (for review: Terry et al., 1994). Loss of cholinergic nerve cells leads to decreased ACh-levels, a loss of cholinergic functions and, consequently, loss of cognition and memory. Despite an immense effort, the causes of nerve cell death in AD are not fully understood. On the basis of particular biochemical, morphological and cellular characteristics, which will be pointed out here shortly, the pathways of cell death may be divided into **necrosis** and **apoptosis**. The process of apoptosis has been mentioned before with respect to regenerative events in the uterus and the bone. Apoptosis occurs throughout the body and is frequently called programmed cell death (PCD). The term PCD suggests the intriguing possibility that such a "death program" may, perhaps, be interrupted in order to stop neurodegeneration. While apoptosis is frequently used in the description of pathophysiological conditions, the term PCD often refers to the physiological way of apoptotic cell death. In the literature very often the terms *apoptosis* and *PCD* are used synonymously. Despite the existing clear-cut definitions there are various overlaps between necrotic and apoptotic cell death and eventually a cell in which an apoptotic pathway has been initiated finally dies displaying necrotic features. In order to find ways to prevent or interrupt nerve cell death one has to understand the mechanistic differences between apoptosis and necrosis.

The two main routes to cell death: apoptosis and necrosis

Apoptosis and necrosis are mechanisms that are not restricted to nerve cells. Most of the knowledge about apoptosis is derived from studies on the immune system. The major characteristics of apoptosis of immune cells also apply to PCD in other target tissue, including the brain. Apoptosis and its physiological counterpart, PCD, play a central role during embryogenesis, fetal development, and immune cell maturation as well as homeostasis.

> Apoptosis/PCD is a essential physiological mechanisms that controls the number of cells in various tissues during development of the organism.

This may best be explained using an example. During the maturation and development of the mammalian nervous system nerve cells are generated in great excess to provide the substrate for the establishment and intact wiring of the nervous tissue. But there is a huge overload of the number of nerve cells. These extra cells are removed via PCD. PCD is not occurring randomly but is rather tightly controlled by various genes. It can be easily imagined that the genetic dysregulation of apoptosis/PCD leads to an inappropriate number of cells and may, ultimately, lead to the development of further uncontrolled cell proliferation and cancer (for review: Wyllie et al., 1980; Blagosklonny, 1999). Dysregulation of apoptosis is discussed in the context of various human diseases such as:

- AIDS,
- autoimmune disease,
- cancer,
- various pathologies in the CNS.

Apoptosis is a process tightly controlled by genes. This is a significant observation. The cells have an intrinsic self-destruction program that is controlled by genes, so it could be said that cells commit suicide. Apoptotic genes can be physiologically and pathophysiologically activated by environmental stimuli including DNA damage, oxidative stress, exposure to hormones, drugs, toxins, virus, and withdrawal of trophic support. As mentioned, apoptosis is a characteristic feature for the development of the mammalian nervous system. A central paradigm of apoptotic/programmed cell death can be easily observed in the nervous system of vertebrates (for review: Burek and Oppenheim, 1996). An estimated number of about 50% of post-mitotic neurons do not receive enough trophic support and die during embryogenesis or fetal development. Its the lack of trophic support that causes physiological apoptotic cell death. The "survival of the fittest" cells, which means here that those cells survive that receive enough trophic signals. Therefore, the major apoptosis trigger for this targeted removal of excess neurons is the competition of too many cells for a limited amount of trophic support provided by for instance NGF. During development excess neurons starve to death (for review: Raff et al., 1993).

In contrast, necrosis is not controlled by genetic programs and can be characterized by a passive pathological event, which frequently is a spontaneous insult or trauma for the cells. This main difference between apoptosis and necrosis is also documented by the different pathological features of these two basic cell death mechanisms. Necrosis is characterized by:

- cellular edema,
- mitochondrial swelling,
- nuclear pyknosis of multiple cells.

Consequently, the plasma membrane disintegrates and the lysis of the cell releases modulators (e.g. enzymes, glutamate) that causes a massive inflammatory response. In the tissue phagocytosis by macrophages clears off the fragments of the cell death. Therefore, necrosis in tissue is always accompanied by inflammation.

Apoptosis, on the other hand, is characterized by:

- cell shrinkage,
- blebbing of membranes,
- nuclear chromatin condensation.

The disintegration of the apoptotic cell does not occur spontaneously but rather the apoptotic process leads to cell fragmentation into small so-called apoptotic bodies. The morphology of the mitochondria as well as of most other organelles is not altered. The apoptotic bodies are then immediately phagocytozed by neighboring cells and macrophages which avoids an inflammatory response in the apoptotic tissue *in vivo* (for review: Williams and Smith, 1993).

Executioners of apoptosis: caspases

It was already mentioned that apoptosis follows a genetic program and a defined sequence of biochemical events. The biochemistry of apoptosis is executed by a family of cysteine proteases called *caspases*.

> Caspases form a rather large family of enzymes. Up to date at least twelve members of this family are known and many of them exist as inactive precursors. Cleavage of the inactive precursor activates the enzyme. Caspases are intertwined in an activation network in the cytoplasma of the cell. With respect to nerve cell death one main player in apoptotic cell death is caspase-3.

The importance of caspase-3 is documented by the fact that a caspase-3 knockout mouse develops a dramatic hypertrophy of the brain due to an excess of neurons (Yang et al., 1996). Because caspases are enzymes and enzymes are catalysts of biochemical reactions, therefore, caspases can be regarded as the catalysts, the promotors of cell death. But how do this cell

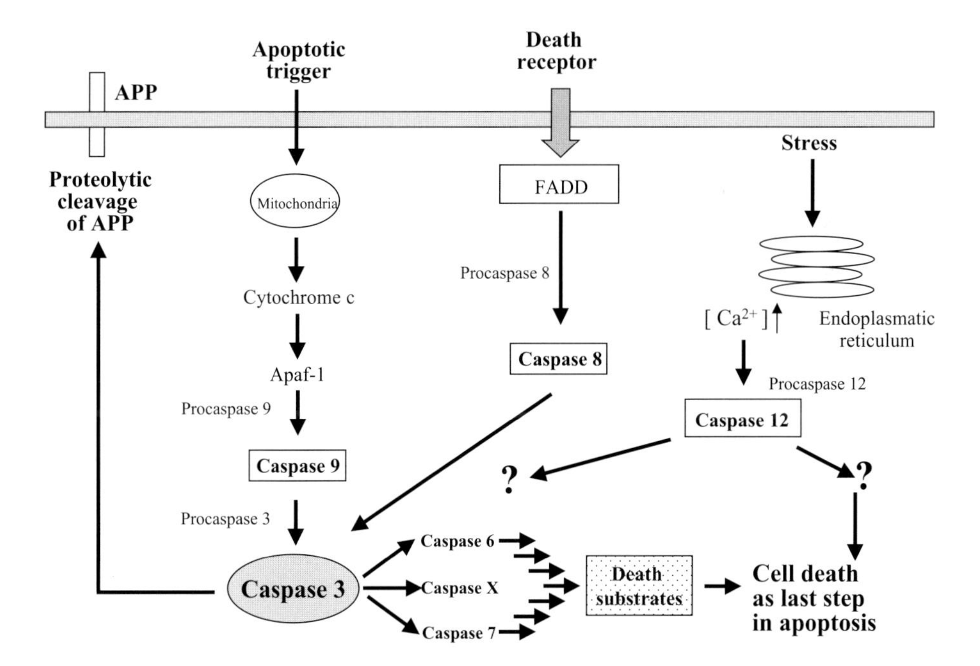

Fig. 39. The caspase network.
Apoptotic triggers that affect mitochondrial function, the activation of death receptors (e.g. Fas/CD95), and intracellular stress can initiate the caspase cascade of apoptosis. The activation of caspase 3 is a common step in this chain of events. Death substrates include cytoskeletal protein and DNA and cause, ultimately the dysfunction and death of cells

death executors actually cause cell death and kill the cell? Enzymes need substrates for their activity and many different cellular substrates for caspases have been described but only for a few the relationship of their cleavage by caspases to cell death is really understood. Most importantly, caspases modify and, therefore, inactivate vital cellular proteins that protect living cells from apoptosis. One prominent example is the cleavage of ICAD, the inhibitor of caspase-activated deoxyribonuclease (CAD or DFF45) by caspases. The DNA, the central substrate of life, is degraded by CAD during apoptosis. The destruction of the genetic information, indispensable for cell survival, kills the cell.

Anti-apoptotic proteins such as members of the Bcl-2-family are also substrates for caspases (e.g. Adams et al., 1998; for review: Thornberry and Lazebnik, 1998). The gene for the Bcl-2 protein is a proto-oncogene which was initially described as a gene activated by chromosome translocation in human B-cell lymphomas (bcl). The Bcl-2 protein is an important anti-apoptotic protein which has manifold functions in the survival of nerve and other cells and will be discussed later with respect to possible interactions with estrogen.

Caspases may also directly alter the architecture and structure of the cell by the cleavage of proteins that contribute to the function of the cytoskeleton. Since caspases indeed attack central vital structures of the cell, their detrimental mode of action has been described to be "reminiscent of a well-planned and executed military operation" (Thornberry and Lazebnik, 1998). With respect to cellular protection against apoptosis one obvious approach would be to block the activity of central caspases. However, this is rather difficult since caspases are organized in a fine-balanced network (see Fig. 39). Certain caspases, such as caspase-3, which are central regulators that integrate various caspase pathways may be appropriate pharmaceutical targets. Indeed, the most prominent and most commonly used caspase inhibitor is a peptide inhibitor of caspase-3 called DEVD and cell permeable analogs thereof such as DEVD-fmk. This caspase-3 inhibitor can prevent apoptosis following neurotrophin deprivation *in vitro* and cell death during ischemia in the brain (for review: Miller, 1997). Since the first discovery of caspases various approaches have been targeted towards the inhibition of these enzymes for neuroprotection. Physiological inhibitors of apoptosis (IAP) have been identified in addition to the above mentioned peptidergic IAPs. The potency and applicability of IAPs *in vivo* is still under intensive investigation (for review: Deveraux and Reed, 1999). However, one has to keep in mind that for the use of caspase inhibitors to block nerve cell death, a precise tissue targeting is extremely important. The prevention of cell death in tumor cells would have fatal consequences.

How to detect apoptosis?

A rather technical but nevertheless very important issue is the question of how to detect apoptosis and how to discriminate between apoptosis and necrosis *in vitro* and *in situ*. Frequently, biochemical methods are used to define key events of apoptosis, such as the fragmentation of the cellular DNA into oligonucleosomal fragments.

> Although many exceptions may be found, there is still a certain agreement in the literature that the intranucleosomal DNA fragmentation is one "hallmark" of apoptosis.

To detect the apoptosis-associated degradation of the DNA ethidium bromide-stained agarose gels are used. A typical pattern of DNA degradation during apoptosis can be induced *in vitro*, for instance as presented by Mesner et al. employing rat pheochromocytoma PC12 cells which were subjected to NGF deprivation, which is a well-acknowledged paradigm of induced apoptosis (Mesner et al., 1992). Agarose gel electrophoresis resolves DNA fragments that occur as laddering pattern, where each band is separated by approximately 200 base pairs. In stark contrast, during necrosis a so-called DNA-smear indicative of randomly degraded DNA can be found (for review: Wyllie and Kerr, 1980) (Fig. 40).

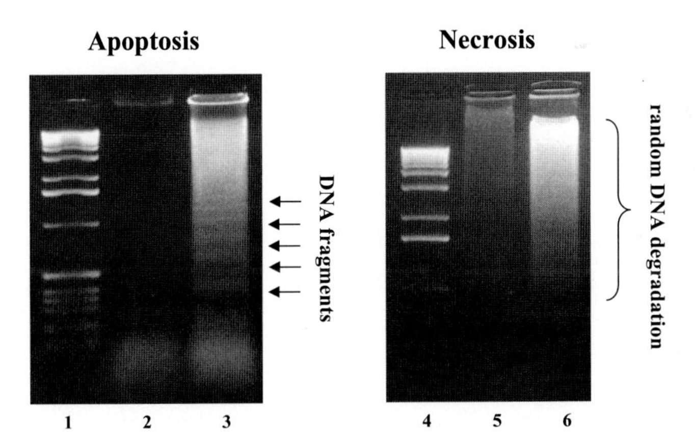

Fig. 40. DNA-laddering as experimental hallmark of apoptosis.
Removal of growth factors from cultured PC12 cells induces apoptosis. The extracted DNA applied to agarose gels give the typical DNA pattern where each band is separated by approximately 200 base pairs (lane 3). In contrast, in necrosis (here induced in PC12 cells by a mitochondrial toxin) a so-called DNA-smear indicative of randomly degraded DNA can be identified in the gel (lane 6). Lane 1 and 4 are DNA-molecular weight markers and lane 2 and 5 depict the untreated controls (no DNA degradation in untreated healthy cells). The agarose gels were stained with ethidium bromide and visualized under UV-light

The pathway of nerve cell death also needs to be detected in tissue, for instance in post mortem tissue. Here, the state-of-the-art method of apoptosis detection in tissue preparations (in situ) is the labeling of the 3′-OH ends of the DNA strand breaks with fluorescein-, or enzyme-conjugated deoxynucleotides, which allows the detection via microscope. Most frequently, the labeling of the DNA breaks is performed using the enzyme terminal transferase and dUTP called terminal deoxynucleotide transferase-(TdT-) mediated dUTP nick-end labeling or TUNEL-labeling (e.g. Sgonc et al., 1994). The break of the DNA opens the nucleotide strands which are then substrates for the enzyme terminal transferase. Labeled nucleotides are attached to the broken strands and can be made visible, e.g. via fluorescence microscopy. It has to be stressed here that the DNA break staining is not entirely specific for apoptosis since in the late phase of necrosis a DNA fragmentation similar to that occurring during apoptosis may also occur, which are then also stained by the TUNEL method. Therefore, a combination of methods need to be applied to safely make the diagnosis of "apoptosis" (for review: Willingham, 1999). One such additional method is the monitoring of the apoptosis-associated chromatin condensation by staining of the

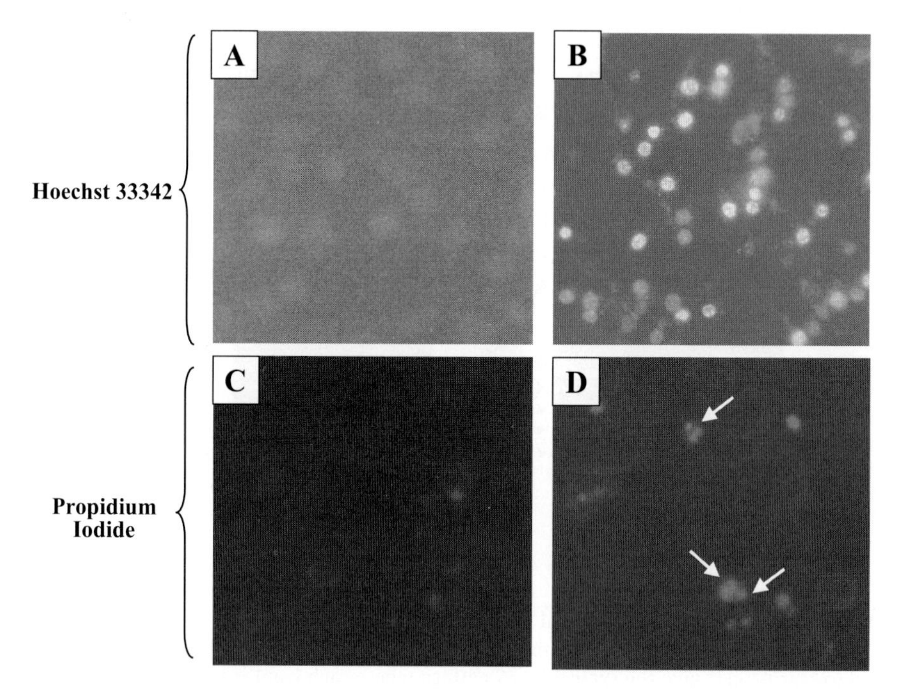

Fig. 41. Staining of apoptotic DNA with a fluorescent dye.
Apoptotic cells in culture can be made visible with fluorescent dyes that interact with the DNA. Staining of the nuclear DNA with Hoechst 33342 increases during apoptotic nuclear condensation (B). Propidium iodide intercalates into the DNA and indicates apoptotic fragmentation (see arrows in D). In A and C untreated control cells are shown

cellular DNA with fluorescence dyes that intercalate into DNA. Frequently used fluorescence dyes are Hoechst 33342 and propidium iodide (Behl et al., 1995). In addition to the various staining procedures of apoptotic tissue in situ a critical microscopical observation of the region of interest will eventually reveal the presence or absence of inflammatory cells (e.g. macrophages) indicative of necrotic processes. All these labeling methods are also frequently used in cell culture systems.

Is there apoptosis in Alzheimer's Disease?

Employing mainly TUNEL labeling and microscopical evaluation apoptosis has been detected not only in AD but also in a variety of other neuropsychiatric disorders including Parkinson's Disease, amyotrophic lateral sclerosis, and stroke. After reviewing the currently available literature that address the question whether apoptosis or necrosis is the predominant feature of nerve cell death in AD, it has to be summarized that the picture is far from being clear.

Apoptosis in post-mortem AD brain tissue

Various histopathological studies detected DNA-fragmentation in tissue sections of brain of AD patients employing the TUNEL technique. Some reports show that the rate of apoptosis in AD brain can be increased up to 50-fold compared to age-matched controls (Yang et al., 1998). Apoptosis was only defined in some of these investigations, by a combination of methods such as TUNEL-labeling and the characterization of the cellular morphology. Some studies reported that despite the positive TUNEL-staining no additional apoptosis features could be found indicating the alternative pathway of cell death, necrosis (Lucassen et al., 1995). Immunohistochemistry also focused on the expression of the well-known executors of apoptosis, the caspases. A comparison of DNA fragmentation with caspase-1 and -3 expression indeed revealed the expression of certain caspases in damaged cells in AD tissue (Masliah et al., 1998). In conclusion, many *in vivo* studies report the presence of dying neurons in the brain of AD patients that display features of apoptosis (for review: Behl, 2000). Many studies exposing cultured cells to suspected triggers of nerve cell death in AD have been used to investigate the molecular pathway of cell death *in vitro*. One important example is the analysis of the pathway of cell death induced by the AD-associated Aβ, which is known to exert neurotoxic activities under certain conditions.

*Apoptosis and necrosis of nerve cells in culture **(in vitro)***

Strong support for the contribution of apoptosis to the overall neurodegeneration in AD comes from *in vitro* studies employing various models of neurotoxicity and also from recent genetic data on AD-linked mutations. AD-associated Aβ develops neurotoxic activities in cultures of primary neurons and of clonal neuronal cells under certain conditions (for review: Yankner, 1996; Behl, 1997). Aβ may indeed induce apoptosis in certain culture models including primary embryonal nerve cells (Loo et al., 1993), but high concentrations of Aβ may also induce nerve cell death with features of necrosis (Behl et al., 1994). This dilemma of conflicting results may partly be due to the fact that Aβ aggregates in cell culture may lead to a first rapid necrosis which is followed by apoptosis throughout the culture. Indeed, it is more than likely that both pathways contribute to Aβ's neurotoxic effects.

All *in vitro* studies of apoptosis employing cultured cells bear an additional complication because apoptotic nerve cell death may be rather cell type-dependent. There is also a concentration dependency since it is known that stimuli that induce apoptosis at low doses can induce necrosis at higher doses. In addition, the discrimination between necrosis and apoptosis may not always be possible since both mechanisms may overlap. Cell death may be initiated by a trigger of necrosis but may finally lead into apoptosis and vice versa. For instance, in certain clonal hippocampal cells (e.g. mouse HT22 cells) oxidative stress induces a form of programmed cell death with the characteristics of both apoptosis and necrosis (for review: Behl, 2000). The lack of phagocytotic cells in *in vitro* systems is a further complication. *In vivo* these cells play a major role in the removal of dead cells and of cellular fragments. This activity is not present when studying mono-type cultures.

AD genetics and apoptosis

Genetics have also given various hints supporting the idea of apoptosis in AD. For instance, the investigation of the pathogenetic role of presenilins in AD also revealed a possible function of these AD-associated genes in AD pathogenesis. Presenilins are proteins that when mutated cause early onset familial AD. The exact physiological function of these transmembrane proteins is not known but missense mutations in the presenilin genes are tightly associated with familial AD and recently a role for presenilin 1 (PS1) and presenilin 2 (PS2) during apoptotic nerve cell death has been suggested (for review: Haass, 1997; Mattson et al., 1998). The transfection and overexpression of the wild-type PS2 in PC12 cells increases the vulnerability of these neuronal cells for apoptosis (Deng et al., 1996). Consistently, the inhibition of PS2 expression employing antisense technique provides protection against apoptotic stimuli. Overexpression of an AD-associated mutated form of PS2 increases the basal apoptotic activity of the cells suggesting that mutated PS2 induces a gain-of-function ultimately leading to neurodegeneration (Wolozin et al., 1996). These data provided the first links between AD-associated familial genes and nerve cell death. Interestingly, this applies also to the gene for PS1. The overexpression of mutated forms of PS1 (but not wild-type PS1) also increases the sensitivity of neuronal cells to apoptosis induced by withdrawal of neurotrophic factors or by neurotoxic stimuli such as Aβ. The modulation of the pro-apoptotic activity of presenilins may be a valuable therapeutic target since both PS1 and PS2 have been shown to also be substrates for caspases. Further genetic evidence supports a role for apoptosis in AD neurodegeneration since also mutated forms of APP may lead to an increased apoptosis in neuronal cell culture and in transgenic mice. Therefore, all three genes that when mutated can cause familial AD have a direct impact on the vulnerability of the cells for apoptosis. It can not be excluded that these genes alter the susceptibility of the cells to necrosis.

What is evident from this discussion is the fact that it is not trivial to describe the exact mechanisms of neuronal cell death *in vitro* and *in vivo*, in general. It is not yet clear what is the exact pathway occurring in AD. Experimental variations and methodological limitations complicate the exact determination of the pathways of cell death. In addition to the various extracellular signals that modulate nerve cell survival and that may cause cell death, certain transcription factor-driven changes may also influence the exact pathway of death. It has to be stressed again, particularly with respect to AD, that although neurodegeneration and progressive nerve cell loss are the key features of AD pathology, it is the loss of synaptic functions, which occurs much earlier in AD pathogenesis. Therefore, it appears that it is the overall stabilization of neuronal function which may be one key to AD prevention and therapy. The lysis of the nerve cell is the final stage of the cellular pathology and its inhibition may be not sufficient to improve or restore neuronal function. Here, the idea has to be raised whether AD is caused by the age-related loss of protection, which step-by-step leads to synaptic dysfunction and, ultimately, to nerve cell death. With respect to AD, certainly one of the protective structures is estrogen since after the loss of estrogen AD incidence increases.

In summary, the analysis of nerve cell death led to the identification of the executors of apoptosis, the caspases. Although in some experimental paradigms of neurodegeneration the inhibition of apoptosis has proven to successfully prevent nerve cell death, it is not very likely that general caspase inhibitors are ideal therapeutic compounds for instance for the treatment of AD. Neurodegenerative disorders can develop during long preclinical phases. Long before apoptosis occurs the dysfunction of synaptic transmission is disturbed. Caspase inhibitors would target cell death but would not restore synaptic function, which would be necessary to counteract the development of AD. On the other hand, caspase inhibitors may be used to prevent sudden cell death as occurring in stroke.

The cause – or very likely the causes – of AD could also be described as a long-term impairment of neuronal function mainly at the synaptic level. The age-related decline in various neurotrophic stimuli may change synaptic plasticity and function and may render neurons more vulnerable to cell death. The accumulation of oxidative stress and the age-related decline in antioxidant defense systems may take part in this scenario. Additional endangering factors that accumulate over time, such as environmental factors or risk factors including head trauma, constantly challenge neurons and may lead to impairment of neuronal function. Considering this, it may be argued that AD is a disease involving a lack of stabilizing and trophic input and a *Loss-of-Protection-Hypothesis* for the pathogenesis of AD can be introduced. Consistent with this view it is well accepted that intrinsic neuronal protective mechanisms decline with age. Consequently, neurons are less well protected. One prominent example is the dramatic drop of the

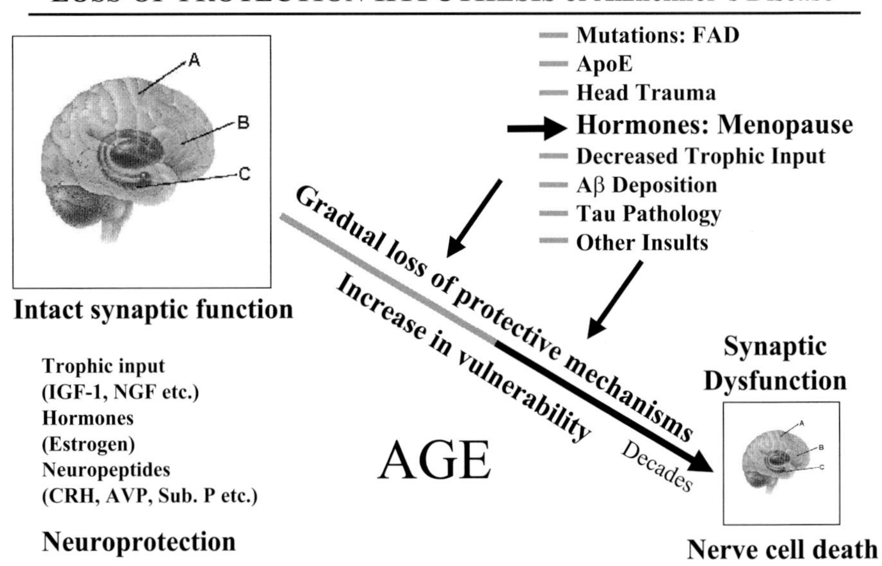

Fig. 42. The loss of protection hypothesis of AD.
The gradual loss of trophic loss and of mediators of neuroprotection including hormones and neuropeptides over decades may render the brain more vulnerable for changes in the local environment and for additional insults which may ultimately lead to the manifestation of neurodegenerative disorders

estrogen level after the menopause that – as argued throughout this book – is associated with an increased incidence of AD in eldery women (for review: Henderson, 2000).

Interestingly, another hormone that shows decreasing levels during aging and AD is the corticotropin-releasing-hormone (CRH), the central modulator of the stress response (for review: Holsboer and Barden, 1996). The CRH-receptor type 1 is expressed in various brain regions and highly in areas of the brain that are not the primary targets of AD-associated neurodegeneration, such as the cerebellum (Behan et al., 1996). We have found that the activation of the CRH-R1 by its ligand CRH protects neurons against oxidative apoptosis (Lezoualc'h et al., 2000). Therefore, it can be suggested that similar to estrogen CRH is a neurohormone displaying intrinsic neuroprotective activities. A loss of these intrinsic protective systems may lead to an enhanced sensitivity of neuronal cells to exogenous insults. Consistent with this view is the finding that CRH has cognition-enhancing effects, just like estrogen and the increase of endogenous levels of CRH has been previously proposed as a potential therapeutic approach for AD (for review: Behan et al., 1996). The demonstration of protective activities of CRH is rather new compared to the detailed knowledge on the neuroprotective actions of estrogen. Data from clinical trials investigating the effect of CRH on AD pathogenesis are not available, so far.

In order to detect cell death and cell protection by estrogen one needs simple and well-established models. It has been already mentioned above how complicated it is to study neuronal cell death and to define the exact pathways of neuronal cell death. Keeping in mind all the intrinsic limitations of cell culture systems, using cultured neuronal cells (clonal as well as freshly prepared primary nerve cells) is still the best way to study molecular intracellular interactions. The first questions which have to be answered here is: How convincing are the cellular data on the protective effects of estrogen? Which models of nerve cell death have been studied so far?

Investigations of estrogen's neuroprotective activities ***in vitro***

Protection of cells by estradiol has been reported in over 10 different neuronal cell types, at least. Those include clonal human, rat, and mouse neuronal cells (e.g. IMR32, PC12, B103; SK-N-SH, SK-N-MC, HT22) as well as freshly prepared cultures of primary hippocampal, neocortical, and mesencephalic neurons. Employing these neruonal cell models various insults have been studied, including the toxicity of Aβ, glutamate and other excitatory amino acids, serum-deprivation, anoxia, hypoxia and others. Interestingly, the majority of these insults induce oxidative stress either directly (e.g. Aβ) or indirectly (e.g. serum-deprivation) at some step of the degenera-

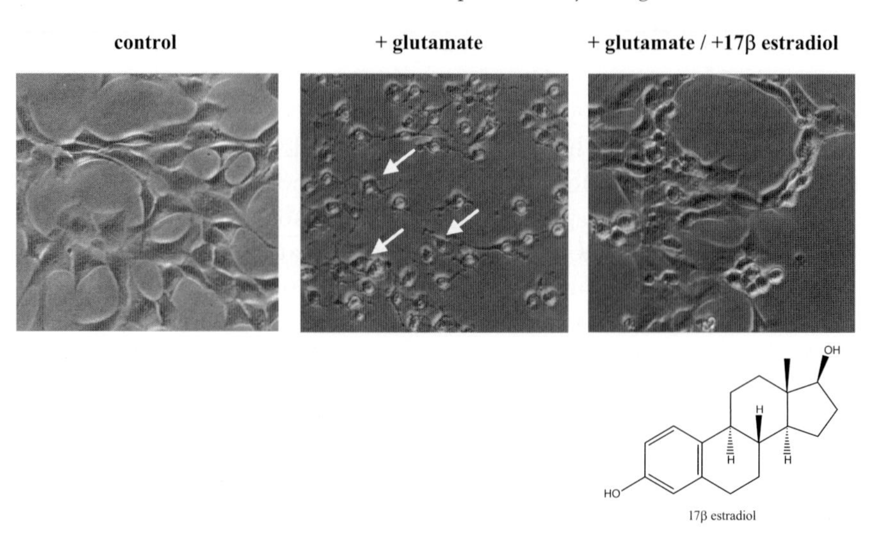

Fig. 43. Antioxidant neuroprotection *in vitro*.
Clonal hippocampal HT22 cells undergoe oxidative cell death upon challenge with glutamate (arrows indicate membrane "blebbing"). 17β-estradiol protects HT22 cells

tion. Therefore, it may be speculated that the structure-related antioxidant activity of estradiol is responsible for the protective effects. In Fig. 43 a typical experiment is shown: clonal hippocampal HT22 cells (Maher and Davis, 1996; Behl et al., 1997; Moosmann and Behl, 1999) were pretreated with 10 μM 17β-estradiol for 12 hours. Then 5 mM glutamate was added to the cell culture to induce oxidative stress. Protection of the HT22 cells against glutamate-induced oxidative apoptosis can be seen by microscopical examination. Of course, in order to determine the extent of cellular protection the extent of cell death occurring in the cell culture needs to be quantified.

Cell survival/cell death can be determined as follows: Most frequently counting of viable cells in the culture is performed. To do so the so-called vital dye trypan-blue is added to the cell culture. This dye enters only those cells with disintegrated membranes. A broken membrane indicates cell death and optical fields were evaluated under the microscope. Stained (dead) cells and non stained (viable) cells are counted and the proportion of living cells per optical field and per total cells is calculated.

Alternatively, colorimetric assays using special dyes and substrates are employed which indicate either the integrity of the cellular metabolism or the integrity of the cell membrane. The cleavage of the compound 3-(4,5-dimethylthiazol-2-yl)-2,5-diphenyltetrazolium bromide (MTT) by intracellular enzymes leads to the formation of blue formazan crystals. This MTT-test is frequently used as a highly sensitive first measure of the

impairment of the cellular metabolism caused by oxidative insults (Hansen et al., 1989). The MTT-test does not indicate cell death but rather metabolic impairment and functional cell damage. Different versions of this MTT assay exist, which differ in the dye that is used.
The most frequent used colorimetric assay for the determination of actual cell lysis is the so-called lactate dehydrogenase (LDH) release assay. Upon the disintegration of the cell the cytoplasmic content is released into the supernatant. LDH is also released and this enzyme can be assayed through its enzymatic activity leading to the enzymatic conversion of a specific substrate in the test mixture. The more LDH is present in the supernatant, the more substrate will be converted to a colored product by LDH indicating of more dead cells.
Alternatively to the MTT- and LDH-assays certain fluorescence dyes are frequently used to determine cell death. Propidium iodide and Hoechst 33342 are dyes that enter cells with disintergated membranes and bind to the DNA in the nucleus. An intensive staining of the nuclei with this dye indicates a large extent of cell death. These dyes have the great advantage that also the morphology of the nucleus is displayed which gives some indication of the type of cell death that occurred, apoptosis versus necrosis (see Fig. 41).

When comparing the different reports about estrogen's potential neuroprotective effects against certain insults it is obvious that there are differences in the actual concentration range in which estradiol mediates this protection *in vitro*. The protective concentrations are in the range from 0.0001 μM (0.1 nM) up to 10 and 50 μM. Most importantly of all, the type of mechanisms (or overlapping mechanisms) that mediate neuroprotection are of central importance. For instance ERs are activated by low nanomolar concentrations. These low concentrations can already induce long-term effects e.g. the transcription of neuroprotective genes. Short-term non-genomic effects, for instance antioxidant neuroprotection, may need high concentrations for several reasons mentioned below. Indeed, a great variability can be observed in the different reports on short-term neuroprotective actions of estradiol *in vitro*. But there are several parameters that need to be considered when judging these obvious discrepancies. With respect to estrogen's antioxidant neuroprotective activity *in vitro* according to the literature I would propose the following general important parameters, which influence the outcome of the cell survival and cell protection assays:

(1) *Cell type*. One central difference is whether primary nerve cells, which are differentiated non-dividing cells or whether clonal (tumor-like) and still dividing cells are used. The expression of certain intracellular modulators of cell survival and players in the antioxidant protection, e.g. co-factors, antioxidant enzymes, may vary, which also affects the potency

of the antioxidant neuroprotection by estradiol. Although there is common sense that the phenolic group of the estradiol molecule is the prerequisite of the purely antioxidant neuroprotection and, therefore, this follows an ER-independent pathway, in certain cellular paradigms ERs may also take part in neuroprotection; and the expression of ERs is quite different depending on the particular cell type. Also, it can not be excluded that ERs when present also take part in neuroprotection. Therefore, cells expressing ERs respond quite differently to estrogen compared to cells that lack ER expression.

(2) *Type of insult*. Oxidative stress can be induced by various compounds and conditions. One example is Aβ. While Aβ induces a a rather delayed oxidative stress at micromolar concentrations L-glutamate can cause oxidative apoptosis within hours. In addition, certain pharmacologically active compounds such as the dopamine D2 receptor antagonist haloperidol induces oxidative damage very rapidly (e.g. Behl et al., 1995; Post et al., 1998). Glutamate can induce neuronal cell death either dependent or independent from glutamate receptor activation (for review: Coyle and Puttfarcken, 1993). Interestingly, both pathways of glutamate-induced cell death converge in the generation of oxygen free radicals and ROS although with a different kinetic. Since the molecule estradiol also underlies chemical degradation in the cell culture, some of the phenol estradiol may be non-specifically oxidized by the time the free radicals need to be scavenged. The

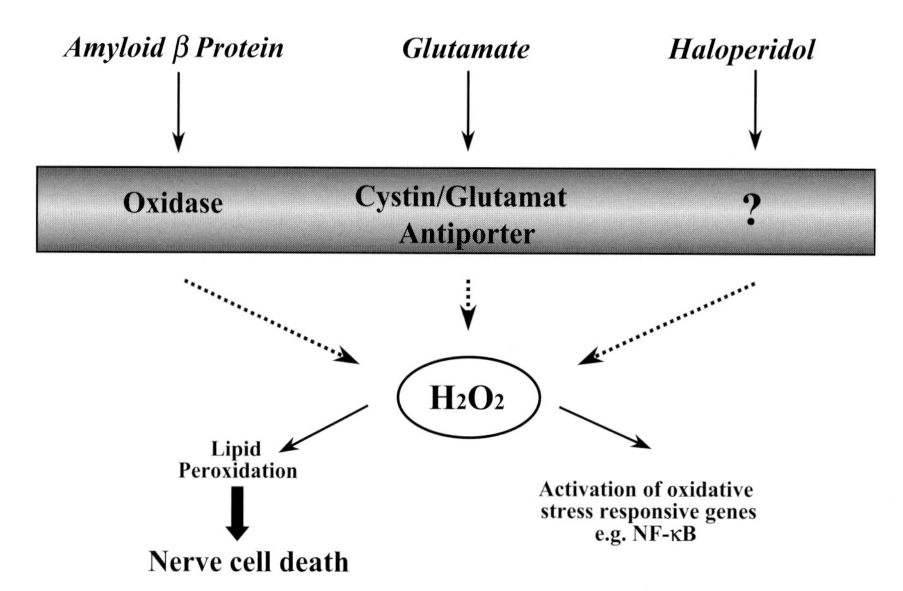

Fig. 44. Final common oxidative pathway induced by various neurotoxins.
Various potential neurotoxins including Aβ and glutamate share a final common oxidative pathway leading to the generation and accumulation of ROS. The activation of redox-sensitive transcription factors such as NF-κB can be induced as part of a stress response of the cell

authors laboratory and others frequently use H_2O_2 to induce oxidative stress in neuronal cells since H_2O_2 imitates the downstream effects of various oxidative neurotoxins, including Aβ and glutamate (Fig. 44). The great advantage of using H_2O_2 is that this compound easily passes the neuronal membrane and its toxic effect is highly reproducible. In most nerve cells H_2O_2 induces oxidative apoptosis.

(3) *Culture conditions.* Cell culture media may vary in the content of various supplements which influence the antioxidant effects. Dependent on the cell type, different media need to be used for propagation of the cells in culture. In order to compare results one has to consider the content of certain supplements and trace elements in the media formulation including antioxidants (e.g. vitamin C), glutathione (GSH), Fe^{2+}-ions, and others. Media that contain protective factors, of course, affect the neuroprotection observed by estradiol. Estradiol may indeed synergistically interact with glutathione present in the culture media. This may decrease the effective concentration of estradiol (Gridley et al., 1998). Although these results are *in vitro* studies numerous possible interactions of estradiol need to be considered when investigating estradiol's protective potential *in vivo*.

(4) *Cell density.* The culture of primary and clonal neuronal cells bears another level of intrinsic variability that is the particular density of the cells used in the survival assays. Cultured cells release certain trophic factors and extracellular matrix molecules, which are essential for their survival, into the supernatant. The growth of clonal cells is highly dependent on the cell number in the culture dish. Cells in a subconfluent cell culture plate have a higher proliferation/doubling rate than cells plated at low density. Proliferating cells can accept much more toxin (e.g. hydrogen peroxide) compared to cells in low density. This also mirrors the need for the protectant. The high density culture may need much more of the estradiol than the low density culture. Although this may not be applicable to all clonal cell cultures it applies to some, which are frequently used for protection studies (e.g. SK-N-MC, HT22). Gene expression in slowly dividing cells can be quite different from those in fast growing cells. When adding high concentrations of H_2O_2 to clonal cells to achieve a detectable level of cell death it is understandable that also high concentrations of protective molecules are needed. For instance when observing antioxidant effects it should be clear that very frequently the free radical interacts with the scavenging molecule (the antioxidant) in a one to one proportion: one antioxidant molecule detoxifies one radical.

(5) *Read-out of cell death.* An important source of variability when determining estrogens effective protective concentrations in certain paradigms is the method used to detect cell survival. Of course, judging the intact morphology of neurons under the microscope can be associated with a rather high individual error. The determination of a dead cell and therefore the

actual cell count may differ from individual experimentor to experimentor. Therefore, a combination of cell survival assays including colorimetric assays which are less subjective, should be entertained for the establishment of dose-response relationships.

For the search of protective lead structures easy-to-read cell-based assays are essential. At the industrial level hundreds of thousands of compounds need to be tested in short time for instance with respect to their antioxidant neuroprotective potential. Therefore, colorimetric assays are ideal for such High-Throughput-Assays.

As mentioned above the concentration range in which estradiol shows protection *in vitro* ranges from nanomolar to micromolar. The physiological relevance of these rather high protective concentrations are still a matter of debate but in most cases are explainable by the design of the corresponding *in vitro* toxicity assays and by considering above mentioned critical variables. The antioxidant neuroprotective activity of estradiol (the antioxidant phenolic lead structure) can be structurally separated from their ER-activating classical properties as hormones. This may lead to the identification of more potent antioxidant compounds (Moosmann and Behl, 1999; see page 180). With respect to the physiology of such estrogen concentrations used *in vitro* compared to *in vivo* estrogen levels one has to take into account:

- the concentration of estrogens is highly variable depending on the menstrual-cycle and may indeed reach lower nanomolar concentrations *in vivo* at certain sites,
- in general, ovarian steroids are concentrated several-fold in the brain relative to plasma and the turnover rate of brain sequestration of blood-borne sex steroids is rather high when compared e.g. with corticosteroids (Pardridge et al., 1980),
- in the blood stream estradiol is bound to large binding proteins and the concentration of free estradiol is indeed very low,
- *in vitro* concentrations may not be extrapolated to the *in vivo* situation since *in vitro* very high concentrations of toxin need to be used (e.g. 1-500 μM hydrogen peroxide) to induce a detectable extent of neuronal cell death,
- the possibility of the accumulation of estradiol at certain sites, for instance in the membrane, needs to be considered as well,
- drugs for effective treatment may reach pharmacological levels rather than physiological concentrations.

In summary, most of the frequently used paradigms of spontaneous direct cellular toxicity in which estradiol increases neuronal cell survival are directly or indirectly associated with oxidative stress. Therefore, most of estradiol's neuroprotective activity could be of non-genomic nature in these paradigms. Next attention is given to the wide range of molecular mechanisms that potentially mediate estradiol's neuroprotective actions including estradiol's antioxidant actions.

Intracellular molecular mechanisms of neuroprotection by estrogen

What we have learned in the previous paragraphs is that estradiol when added to cultivated nerve cells protects these cells against oxidative damage and other insults, which may directly or indirectly cause oxidative stress. No ERs are necessary for this action. But of course, a variety of trophic and protective effects of estrogen are based on the activation of ERs and the resulting changes in transcription. One has to keep in mind that the wide range of possible interactions and cross-talks of estrogen with its receptors, with other signaling pathways and with different cellular structures makes it really difficult to get an integrated picture of neuroprotection caused by estrogen.

Given the scenario that estrogen indeed is present in certain brain regions at a concentration high enough to act as an antioxidant, estradiol can still activate ERs and other signaling pathways in that cell or tissue. The consequence is neuronal gene transcription, which may then also have vital consequences. What genes are induced by estradiol and can actually act as protectants?

Direct ER-dependent neuroprotection: induction of neuroprotective genes

A quick reminder concerning estradiol's action via ERs: estrogen enters the cell via diffusion over the cell membrane and binds to ERs, which then translocate into the nucleus. There, dimers of ERs bind to EREs (directly or via adaptor proteins) and enables the transcriptional machinery to start transcription. This "classical" estrogen activity has been frequently mentioned in the discussion so far. In nerve cells that express functional ERs, which are activated by estrogen, the transcription of a variety of genes is induced. For instance in hypothalamic cells, the traditional target cells of estrogen in the brain, genes which code for proteins that directly or indirectly modulate sexual behavior and reproduction are activated, such as oxytocin. Various genes in the genome carry EREs in their promoter regions.

Several neuronal genes are known that are activated by E2 (via its receptor) that can be called ER target genes. Among those genes are candidates that are known as powerful neurotrophic and neuroprotective players. The gene for the brain derived neurotrophic factor **(BDNF)** contains an ERE and the expression of BDNF is stimulated by E2 (Sohrabji F. et al., 1995;

Oxidative destruction in neurodegenerative disorders may either be a result from a lack of function or from a gain of function of antioxidant enzymes as reported for the motor neuron disease (ALS). In ALS SOD appears to be functionally altered. A genetic link supports this view since in familiar cases of ALS a mutation in the enzyme SOD1 has been found (Rosen et al., 1993). During stroke which is linked to ischemia and reperfusion processes, oxidation reactions may occur as a secondary damaging effect (for review: Love, 1999). During PD the metabolism of dopamine and its chemical breakdown by monoamine oxidases may lead to the formation of ROS, which may damage dopaminergic neurons (for review: Ebadi et al., 1996; Floor, 2000). In the last decade many experimental data based on molecular, cellular, pathological, and *in vivo* studies strongly support an important role for ROS in the pathogenesis and progression of AD (for review: Bains and Shaw, 1997; Ando et al., 1998; Behl, 1999; Markesbery, 1999; Prasad et al., 2000). Some of the evidence involves the following findings:

- Post mortem evidence for an oxidative burden in AD: Increased levels of oxdidation end products are found in Alzheimer´s tissue compared to age-matched controls, increased oxidation of DNA and of membrane lipids are found.
- Inflammatory processes: due to inflammatory processes in the AD brain accumulating ROS may challenge the surrounding nerve cells.
- Deposits of Aβ induce an overal oxidative environment (see Fig. 34).
- There is an over all decreased age-associated decline of the antioxidant defense that renders the brain more vulnerable.

Estradiol is a neuroprotective antioxidant

Although it has been known for quite some time that phenolic groups can act as antioxidants, the fact that 17β-estradiol is a neuroprotective antioxidant was surprising. Why? Because estrogen is known as a hormone acting via intracellular hormone receptors. Although estrogen's interaction with integral proteins and receptors of the membrane has led to the understanding of estrogen as neuroactive compound, the intrinsic phenolic structure and its potential has not been considered fully.

In many experimental paradigms *in vitro* and *in vivo* of nerve cell death various estrogens and estrogen-derivatives can act as neuroprotective antioxidants. The main structural requirement is the basic phenolic structure of the molecules (Behl et al., 1995, 1997; Green et al., 1997; Dubal et al., 1998; Wang et al., 1999). In stark contrast to above mentioned interactions of E2 with ER and, in part, also with intracellular signaling pathways, the antioxidant effect of estradiol is completely independent from the presence and activation of ERs. This has been demonstrated by a variety of groups. Estradiol and some estradiol derivatives act as antioxidants also in cell-free test tube assays (e.g. Behl et al., 1995; Green et al., 1997; Moosmann and Behl, 1999).

Of course, the most prominent natural lipophilic phenolic antioxidant is vitamin E (α-tocopherol). Vitamin E is a monophenolic compound and acts as lipophilic antioxidant in a variety of models. With respect to neurodegeneration, vitamin E prevents the accumulation of oxidative metabolites induced by Aβ as well as Aβ's neurotoxicity *in vitro* (Behl et al., 1992, 1994; Harris et al., 1995). Based on this and other evidence, vitamin E has been recently used in clinical trials for the treatment of AD. In one first multicenter study employing high doses (2000 international units, I.U., daily) of vitamin E in moderately severe Alzheimer patients, some beneficial effects of the use of vitamin E have been reported (Sano et al., 1997).

Estradiol is a monophenolic compound just like vitamin E. The direct comparison of the two structures reveals that both molecules consist of a phenolic antioxidant acting moiety and of a lipophilic moiety (see Fig. 22). By modifying estrogen's particular molecular structure by using chemical estrogen-derivatives and other phenolic compounds in paradigms of oxidative nerve cell death, it was revealed that, in general, aromatic alcohols or phenolic compounds exert antioxidant neuroprotective activities. It has to be given that some minimal accessory prerequisites are present such as the potential of the compound to penetrate membranes and a high degree of lipophilicity (Moosmann et al., 1997). These latter prerequisites can be explained on the basis of the idea that early hydrophilic lipid peroxidation breakdown products may be causative for damage to aqueous phase proteins and nucleic acids (Esterbauer et al., 1991).

Interestingly, compounds such as serotonin (5-hydroxytryptamine), normelatonin (N-acetyl-5-hydroxytryptamine, Fig. 1), quercetin, and also simple alkylphenols including 4-dodecylphenol and trimethylphenol (Fig. 48) all protect neuronal cells against oxidative cell death *in vitro* and prevent membrane lipid peroxidation (Moosmann et al., 1997; Skaper et al., 1997). The impact of a phenol group on the antioxidant potential of a compound is clearly demonstrated by the direct comparison of melatonin (N-acetyl-5-methoxytryptamine) and its molecular precursor normelatonin (N-acetyl-5-hydroxytryptamine). Here, the presence of a phenolic group, such as in normelatonin, is enhancing the direct antioxidant activity. Indeed, melatonin is not a phenol and part of its effect has to be assigned to another group in the molecule (the indol) and, perhaps, to melatonin's activity at melatonin receptors (Moosmann et al., 1997; Lezoualc'h et al., 1998; for review: Reiter, 1998).

The chemical family of phenolic compounds is huge and some of these structures indeed have the power to prevent oxidation in neurodegenerative paradigms. Recently, other members of this phenol-family have been demonstrated to exert neuroprotective activities. They belong to the well-know psychoactive **cannabinoids**. Cannabinoids with potent antioxidant neuroprotective activity *in vitro* include cannabidiol and (-)Δ^9-tetra-hydro-cannabinol (THC, Fig. 48). THC was already known for its activity in preventing hydrogenperoxide-induced oxidative damage in a chemical system such as during the Fenton reaction and as well as in neuronal cultures. In addition, cannabidiol is neuroprotective against glutamate neurotoxicity and interestingly more effective than either ascorbate (vitamin C) or α-tocopherol, underlining its potential to act as antioxidant (Hampson et al., 1998).

Another well-acknowledged group of phenolic anitoxidants are the **flavonoids** (Tereao et al., 1994; Fauconneau et al., 1997; Lien et al., 1999). Unfortunately, comparative data on their antioxidant neuroprotective effects are lacking. Nevertheless, interesting *in vitro* effects have been found, e.g. for quercetin (Oyama et al., 1994; Skaper et al., 1997; Moosmann and Behl, 1999; for review: Moosmann and Behl, 2000).

Coming back to the monophenolic molecule estradiol. Estrogens including 17β-estradiol, its stereoisomer 17α-estradiol, and ethinyl estradiol, protect cultured neurons against oxidative cell death, which is – as stressed already – characteristic for AD and PD. The antioxidant lead structure is the

α-tocopherol

2,4,6-trimethylphenol

quercetin

17β-estradiol

Delta 9-Tetrahydrocannabinol

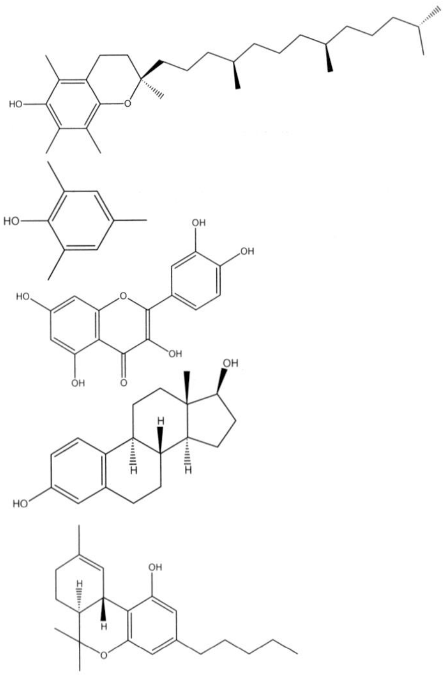

Fig. 48. Neuroprotective antioxidants.
Common to this antioxidants that prevent oxidative apoptosis *in vitro* is the presence of the hydroxyl group at a mesomeric ring system

hydroxyl group in the so-called ring A of the steroid structure. Chemical modification of this particular moiety such as through etherization (e.g. mestranol, methyl ether of ethinyl estradiol) completely blocks the antioxidant activity of the molecule.

In general, phenolic A ring estrogens are powerful inhibitors of lipid peroxidations in various cell free test-tube models (Nakano et al., 1987; Sugioka et al., 1987; Hall et al., 1991; Mooradian, 1993; Lacort et al., 1995; Moosmann and Behl, 1999). Initially it was described that estradiol generally increases survival of cultivated clonal glial and neuronal cells (Bishop and Simpkins, 1994). Soon after, estradiols protective effect in paradigms of oxidative stress induced by Aβ, H_2O_2 and glutamate was shown and a direct antioxidant effect that is independent of ER activation was demonstrated (Behl et al., 1995). Since then many laboratories confirmed this antioxidant activity in their particular experimental paradigm of oxidative cell degeneration (Behl et al., 1995, 1997; Goodman et al., 1996; Green et al., 1997; Blum-Degen et al., 1998; Sagara et al., 1998; Lezoualc'h and Behl, 1998; for review: Behl, 1999; Green and Simpkins, 2000). 17β-estradiol can prevent cellular oxidations and cell death as induced by:

- Aβ,
- glutamate,
- $FeSO_4$,
- haloperidol,
- H_2O_2.

Antioxidant effects are reached in most *in vitro* studies at a concentration range of 1–10 μM. One paradigm demonstrated protective effects against oxidative neuronal cell death within nanomolar concentrations (Gridley et al., 1997), which has been, in part, attributed to possible enhancing effects induced by the presence of GSH in the culture media (Gridley et al., 1998). The discussion concerning the minimal effective dose of E2 needed for the protection of cultured cells against oxidations is very hard to follow since differences in the formular of media and sera used for cell culture may affect the outcome of the experiment (see page 159). Interactions and redox cyclings such as that occurring between α-tocopherol and ascorbate may significantly account for lowering of effective E2 concentrations. Leaving aside the discussion of how much estrogen is necessary to reach a significant antioxidant protective effect in *in vitro* models, it is quite obvious that the *in vitro* concentrations are higher than the levels normally reached in the body. And therefore, the physiological relevance of these (nanomolar to micromolar) concentrations needs to be discussed. Concerning this point one has to take into account: (1) to reach a pronounced and well detectable oxidative damage and cell death one has to use rather high concentrations of oxidants *in vitro*, e.g. mM H_2O_2, and, consequently, high concentrations of antioxidant molecules are necessary, (2) the concentration of estrogens is highly variable and depend on the menstrual-cycle and, in some cases, may indeed reach lower nanomolar concentrations *in vivo*, (3) in general, ovarian steroids are concentrated several-fold in brain relative to blood plasma and it can not be determined exactly how much estrogen accumulated at or in the cell membrane, (4) the turnover rate of brain sequestration of blood-borne sex steroids is rather high compared for instance with corticosteroids (Pardridge et al., 1980), (5) drugs for effective prevention of damage or treatment may reach pharmacological levels rather than physiological concentrations.

It is the antioxidant effect of estrogen based on the potential of a phenolic compound to interact with free radicals that is of great interest. Given the importance of free radicals in the regulation of physiological processes, for instance activation of redox-sensitive transcription factors (NF-κB, AP-1), and considering the vital role of the intracellular redox mileu for cell survival, this activity of estrogen definitively needs to be taken into account. The approach to design antioxidant molecules which may be used *in vivo* and, ultimately, in clincial trials is a completely different issue. The authors group is consequently following this concept. In the search for compounds that are potent phenolic antioxidants but have no estrogenic, meaning ER activating, properties, the small molecule 2,4,6-trimethylphenol (TMP) was identified (Fig. 48). TMP does not bind to the ER and does not activate ER-dependent gene transcription but, nevertheless, is a potent antioxidant preventing lipid peroxidation and oxidative neuronal cell death (Moosmann and Behl, 1999). TMP is a rather small compound and is very likely easily passing the blood brain barrier. Therefore, this compound is currently being tested for its protective potential *in vivo* employing an established stroke model (global ischemia in the Mongolian Gerbil). The first experimental data are very promising and demonstrate a powerful antioxidant and cytoprotective effect also *in vivo*; here, the most vulnerable hippocampal CA1 region was found to be highly protected (Behl, unpublished). Ideally, such non-estrogenic compounds could be applied in both male and female organisms since hormonal estrogenic side effects are excluded. As mentioned (see page 128) 17β-estradiol has also potential to act as protectant in stroke models.

The antioxidant potential of the phenolic structure of estradiol may add to the overall beneficial and protective effects during neurodegeneration. Of course, among the multiple neuromodulatory roles *neuroprotection* through estrogens is of increasing interest. This interest is fueled in part by the already known beneficial effects of ERT in the prevention and treatment of age-related physiological changes, in general, and neurodegenerative diseases such as AD, in particular. It has to be stressed that up to now it has not really been proven what particular function of estrogens (receptor-dependent versus receptor-independent, modulation of intracellular signaling, structural effects) is mediating the manifold neuroprotective activities in various disease-states and in normal brain functions. This holds true also for the well-known gender differences of the incidence of many neuronal disorders. The data accumulated so far underline that the steroid hormone estrogen, the lipophilic phenolic compound 17β-estradiol, exerts neuroprotective effects at various cellular levels: the membrane (e.g. by modulating neurotransmission), the cytoplasm (e.g. by acting as free radical scavenger and by cross-talking with intracellular signaling), and the nucleus (via the activation of nuclear ERs). The use of compounds such as SERMs that specifically modulate ER action or of certain estradiol-derivatives lacking estrogenic function in certain paradigms will reveal what estrogen-like structure is best in which paradigm of nerve cell death. Of big importance is also the proof-of-concept *in vivo* and in large clinical trials. In addition, the identification of non-estrogenic lead structures with a high antioxidant potential and good pharmacological characteristics may lead to the design of a "super-antioxidant".

It is very likely that many of the beneficial long-term effects seen in ERT with respect to various diseases are mediated via novel genetic programs induced by estrogen. Such estrogen-target genes may represent important neuroprotective targets. Ideally, such candidate genes are studied with respect to the exact structure of the gene and its inducibility and then external modes of pharmacological activation are designed. This would mean **estrogen lays the trail towards neuroprotective gene expression** which then is induced independently by other drugs. State-of-the art methods exist that can be used to identify the potential neuroprotective expression profile induced by estrogen.

Gene chips were introduced by the company Affymetrix Inc. (Lipshutz et al., 1995; Chee et al., 1996; Hacia et al., 1996), which is using a special experimental trick to generate such DNA chips. The glass slide of the chip is able to bind the oligonucleotides A, T, G, or C. The oligonucleotides are synthesized step-by-step using the special sequence of masks that overlay the chip and allow the conjugation of certain nucleotides at certain spots. The Affymetrix technology allows to synthesize 10 000 to 1 000 000 different oligonucleotides on one chip. The nucleotides are coupled via photolithography and the sequential use of unique masks (for review: Watson and Akil, 1999; Celis et al., 2000; Young, 2000).

Screening DNA-arrays/gene chips to identify estrogen target genes

The use of DNA-arrays and gene chips should be explained by an experimental example: for hybridization to the DNA array carrying the ESTs or the gene chip carrying oligonucleotides, the mRNAs from the e.g. brain tissue of estrogen-treated animals and those of the non-treated control animals are isolated. Next, the mRNAs are transcribed into cDNAs and in this transcription step a fluorescent label is incorporated into the cDNA sequence. Then the two populations of fluorescent-labeled mRNAs/cDNAs are hybridized with the DNA arrays/gene chips and in separate sets of hybridization the binding pattern of the two different mRNA/cDNA populations can be compared. The pattern is recorded using laser confocal microsopy. The signals over each of the locations are detected, recorded, and quantified for each spot on the array. When using different fluorescent dyes (green label for one mRNA/cDNA population and red label for the other one) one can perform the hybridizations in parallel. By the use of laser scanning technology the different labels are recorded and analyzed separately. The expression profiles of both sets (green and red) can be directly compared. To come back to our example the expression profile for the estrogen-treated versus the non-treated control tissue can be directly compared (Fig. 49). The further bioinformatic analysis will exactly identify the genes that are up- or down-regulated in their expression. Of course, one would expect that certain well-known genes carrying EREs are among those identified in the array-analysis/chip screen. In addition, the whole pattern of estrogen-sensitive genes will be identified. Certain estrogen target genes or the combination of genes could then be further investigated for their potential to mediate neuroprotection. Ideally, *master genes of neuroprotection* are detected which then will be studied with respect to their external inducibility. Perhaps, such genes carry well-known response elements in their promoters (not necessarily only EREs) which would enable one to induce its expression e.g. by certain drugs. In industrial measures, the promoters of such neuroprotective

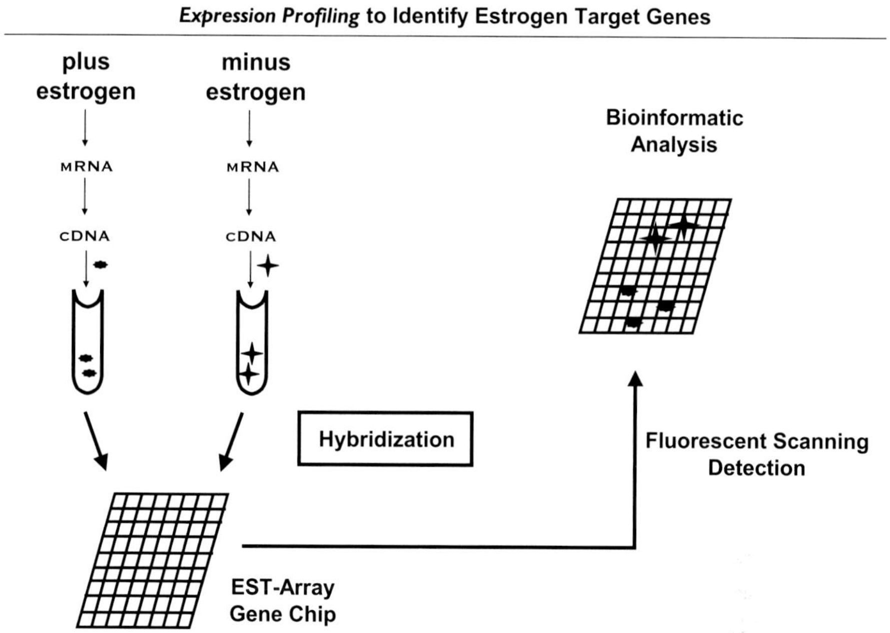

***Expression Profiling* to Identify Estrogen Target Genes**

plus estrogen
minus estrogen
MRNA
MRNA
cDNA
cDNA
Bioinformatic Analysis
Hybridization
Fluorescent Scanning Detection
EST-Array Gene Chip

Fig. 49. Gene Chips.
Gene chips can be employed to establish the expression profiles of cells and tissues. For instance in the search for estrogen-regulated neuronal genes the mRNA of cells (estrogen-treated versus controls) is isolated and transcribed into cDNA. A fluorescent label is introduced during cDNA synthesis. The cDNAs can be hybridized to ESTarrays/gene chips and complementary binding can occur. Fluorescent spots indicative of complementary binding are identified and analyzed with a fluorescent imager. Custom-made gene chips with an array of known genes may identify these genes that are expressed in the sample of interest. Further, a detailed bioinformatic analysis is necessary to effectively handle the acquired data

genes are subcloned into luciferase vectors or into other detection constructs and then *High Throughput Assays* would further analyze this "screenable target gene".

In summary, this new technology has the power to identify the genes whose expression is induced by estrogen. Those genes may include
(1) genes, that are directly regulated by ERs,
(2) genes that are indirectly controlled via other signaling mechanisms, and
(3) genes that are regulated via a second or third regulatory step.
In addition, these expression profilings reveal groups of related genes that are controlled by estrogens. Estrogen, which is a powerful neuroprotectant via various mechanisms shows a trail which is worth following in the identification of potent neuroprotective genes. The vision would then be to design a drug that specifically and selectively induces

that is massively affected during AD has proliferating zones and a neuronal stem cell potential. Again, the stimulation of neuronal stem cell activity and of the controlled differentiation is of central importance. And here, the female sex hormone comes into picture. Various brain areas exert sex specific differences (see page 68). Those brain areas include also the hippocampus. Many of the observed differences in hippocampal structure and function depend on the levels of the gonadal hormones in the body. With respect to neural stem cells it has been shown that estrogen makes also an important difference. To analyze a possible sex difference in the generation of hippocampal cells in the adult brain, proliferating cells and their immediate progeny have been investigated again using BrDU-labeling (Tanapat et al., 1999). And indeed, the removal of the ovaries (ovariectomy) of the adult female rat significantly reduced the number of BrDU-labeled cells. This effect could be reversed by the supplementation of the ovariectomized animal with 17β-estradiol (estrogen replacement).

In summary, these authors found an estrogen-enhanced cell proliferation which occurs during proestrus (when the estrogen levels are at a peak), which results in more immature hippocampal neurons in females when compared to male counterparts. Clearly, females produce more new neuronal cells than males and that this effect is caused by estrogen. Stem cells will be further investigated in the future. Indeed, the use of neuronal stem cells in the therapy of neurodegenerative disorders is a completely new concept and may proof to be a powerful therapeutic avenue.

Final remarks: neuroprotection by estrogen

It is now well accepted that the female sex hormone estrogen has many actions throughout the body extending its role above being a pure *female sex hormone*. Considering

- "female"

Estrogen it is not restricted to the female organism since the male sex hormone testosterone (and other steroid structures with a 19 carbon atom structure, so-called C-19 steroids) is locally converted to estradiol in various tissues including the brain by an aromatase P-450 enzyme.

- "sex"

Indeed, the main source of estrogens are the female ovaries and the traditional activities of estrogen are its important roles during sex maturation and differentiation. But the activity of estrogen goes far beyond the modulation of sex-related processes since multiple target sites for estrogen activity exist, including the brain.

- "hormone"

The classical definition of a hormone does apply to various estrogen effects but definitively not to all since estrogen may act in a hormone receptor-independent (non-genomic) manner.

The knowledge about estrogen and the proteins that mediate many of its activities inside the cell, the ERs, tremendously increased during the last decades. The sciencitifc progress was mainly driven by the awareness that estrogen is the prototype of a sex hormone and it was always seen as a key for female-specific body functions, female sex-specific differences, and female beauty.With the advent of modern cellular and molecular biology in the last twenty years and even more now as large global projects determine the exact DNA sequence of the humane genome (Human Genome Project), much more will be known about estrogen's target genes in various tissues. Modern technologies provided the tools to take a closer look at the neuromodulatory compound estrogen and to define novel activities and novel structural features that are important issues to dissect and understand the full picture of estrogen's activity.

In summary, sex steroids in general and estrogen in particular affect many neuronal and non-neuronal conditions such as mood changes, schizophrenia, neurodegenerative diseases, arteriosclerosis, and osteoporosis. In the nervous system, and in particular in the brain, estrogen is a multifunctional player. Much more research is required in order to establish the precise pathways of estrogen actions in the brain. With respect to neurodegenerative disorders not only estrogen's direct effects on nerve cell survival is of great importance but also the downstream effects on gene expression in nerve cells. Although the outcome of many more clinical trials needs to be awaited, ERT may offer an important new avenue to counteract the increasing incidence of age-related mental disorders and to improve the degenerative symptoms during aging.

10. References

Adams JM, Cory S (1998) The BCL-2 protein family – arbiters of cell survival. Science 281: 1322–1326

Aguzzi A, Raeber AJ (1998) Transgenic models of neurodegeneration. Neurodegeneration: of (transgenic) mice and men. Brain Pathol 8: 695–7

Altman J (1962) Science 135: 1127

Altman J, Das GD (1965) J Comp Neurol 124: 319

Alzheimer A (1907) Über eine eigenartige Erkrankung der Hirnrinde. Allg Zeitschr Psychiat Psych-Gericht Med 64: 146–148

Amaducci LA, Fratiglioni L, Rocca WA, Fieschi C, Livrea P, Pedone D, Bracco L, Lippi A, Gandolfo C, Bino G et al (1986) Risk factors for clinically diagnosed Alzheimer's disease: a case-control study of an Italian population. Neurol 36: 922–31

Ames BN (1983) Dietary carcinogens and anticarcinogens. Oxygen radicals and degenerative diseases. Science 221: 1256–1264

Anand A, Charney DS, Delgado PL, McDougle CJ, Heninger GR, Price LH (1994) Neuroendocrine and behavioral responses to intravenous m-chlorophenylpiperazine (mCPP) in depressed patients and healthy comparison subjects. Am J Psychiatry 151: 1626–30

Ando Y, Suhr O, Elsalhy M (1998) Oxidative stress and amyloidosis. Histol Histopathol 13: 845–850

Anonymous (1993) Effects of tocopherol and deprenyl on the progression of disability in early Parkinson's disease. The Parkinson Study Group. N Engl J Med 328: 176–83

Anstead GM, Carlson KE, Katzenellenbogen JA (1997) The estradiol pharmacophore: ligand structure-estrogen receptor binding affinity relationships and a model for the receptor binding site. Steroids 62: 268–303

Archer JS (1999) NAMS/Solvay Resident Essay Award, Relationship between estrogen, serotonin, and depression. Menopause 6: 71–8

Arnold AP, Breedlove SM (1985) Organizational and activational effects of sex steroids on brain and behavior: a reanalysis. Horm Behav 19: 469–98

Aronica SM, Kraus WL, Katzenellenbogen BS (1994) Estrogen action via the cAMP signaling pathway: stimulation of adenylate cyclase and cAMP-regulated gene transcription. Proc Natl Acad Sci USA 91: 8517–21

Aronson MK, Ooi WL, Morgenstern H, Hafner A, Masur D, Crystal H, Frishman WH, Fisher D, Katzman R (1990) Women, myocardial infarction, and dementia in the very old. Behav Neural Biol 40: 1102–1106

Arouma OI, Evans PJ, Kaur H, Sutcliffe L, Halliwell B (1990) An evaluation of the antioxidant and potential pro-oxidant properties of food additives and of trolox C, vitamin E and probucol. Free Radical Res Com 10: 143–157

Auger AP, Tetel MJ, McCarthy MM (2000) Steroid receptor coactivator-1 (SRC-1) mediates the development of sex-specific brain morphology and behavior. Proc Natl Acad Sci USA 97: 7551–7555

Aust SD, Morehouse LA, Thomas, CE (1985) Role of metals in oxygen radical reactions. J Free Radic Biol Med: 3–25

Avis NE, Brambilla D, McKinlay SM, Vass K (1994) A longitudinal analysis of the association between menopause and depression. Results from the Massachusetts Women's Health Study. Ann Epidemiol 4: 214–20

Baeuerle P, Baltimore D (1988) IκB: a specific inhibitor of the NF-κB transcription factor. Science 242: 540–546

Baichwal VR, Baeuerle PA (1997) Activate NF-κB or die? Curr Biol 7: R94–R96

Bains JS, Shaw CA (1997) Neurodegenerative disorders in humans – the role of glutathione in oxidative stress-mediated neuronal death. Brain Res Rev 25: 335–358

Baldwin AS (1996) The NF-κB and IκB proteins: New discoveries and insights. Annu Rev Immunol 14: 649–683

Baraban JM, Fiore RS, Sanghera JS, Paddon HB, Pelech SL (1993) Identification of p42 mitogen-activated protein kinase as a tyrosine kinase substrate activated by maximal electroconvulsive shock in hippocampus. J Neurochem 60: 330–6

Barber PV, Arnold AG, Evans G (1976) Recurrent hormone dependent chorea: effects of oestrogens and progestogens. Clin Endocrinol 5: 291–3

Barrett-Connor E (1997) Sex differences in coronary heart disease. Why are women so superior? The 1995 Ancel Keys Lecture. Circulation 95: 252–64

Baskin DS, Browning JL, Pirozzolo FJ, Korporaal S, Baskin JA, Appel SH (1999) Brain choline acetyltransferase and mental function in Alzheimer disease. Arch Neurol 56: 1121–3

Baulieu EE (1997) Neurosteroids: of the nervous system, by the nervous system, for the nervous system. Recent Prog Horm Res 52: 1–32

Baulieu EE, Kelly PA (1990) Hormones: From molecules to disease. Chapman and Hall, New York and London

Bayer SA (1982) Changes in the total number of dentate granule cells in juvenile and adult rats: a correlated volumetric and 3H-thymidine autoradiographic study. Exp Brain Res 46: 315–23

Bayliss WM, Starling EH (1902) The mechanism of pancreatic secretion. J Physiol 28: 325–353

Beal MF (1995) Aging, energy, and oxidative stress in neurodegenerative disorders. Ann Neurol 38: 357–366

Becker JB (1990) Direct effect of 17beta-estradiol on striatum: sex differences in dopamine release. Synapse 5: 157–64

Becker JB, Beer ME (1986) The influence of estrogen on nigrostriatal dopamine activity: behavioral and neurochemical evidence for both pre- and postsynaptic components. Behav Brain Res 19: 27–33

Beckman KB, Ames BN (1998) The free radical theory of aging matures. Physiol Rev 78: 547–581

Bedard P, Langelier P, Villeneuve A (1977) Oestrogens and extrapyramidal system. Lancet 2: 1367–8

Beg AA, Baltimore D (1996) An essential role for NF-κB in preventing TNF-κ-induced cell death. Science 274: 782–784

Behan DP, Grigoriadis DE, Lovenberg T, Chalmers D, Heinrichs S, Liaw C, De Souza EB (1996) Neurobiology of corticotropin releasing factor (CRH) receptors and CRH-binding protein: implications for the treatment of CNS disorders. Mol Psych 1: 265–277

Behl C (1997) Amyloid beta-protein toxicity and oxidative stress in Alzheimers-disease. Cell Tiss Res 290: 471–480

Behl C (1999) Alzheimer's disease and oxidative stress: implications for novel therapeutic approaches. Prog Neurobiol 57: 301–323

Behl C, Holsboer F (1999) The female sex hormone oestrogen as a neuroprotectant. Trends Pharmacol Sci 20: 441–4

Behl C (2000) Apoptosis and Alzheimer's disease. J Neurol Transm 107: 1325–1344

Behl C, Davies J, Cole GM, Schubert D (1992) Vitamin E protects nerve cells from amyloid β protein toxicity. Biochem Bioph Res Co 186: 944–950

Behl C, Davies JB, Lesley R, Schubert D (1994) Hydrogen peroxide mediates amyloid β protein toxicity. Cell 77: 817–827

Behl C, Moosmann B (1999) Estrogen and other antioxidants in neuroprotection: relevance to Alzheimer's disease. In: Poli, Packer, Cadenas (eds) Free Radicals in Brain Pathophysiology. Dekker Publications New York, 22: 467–485

Behl C, Skutella T, Lezoualc'h F, Post A, Widmann M, Newton C, Holsboer F (1997) Neuroprotection against oxidative stress by estrogens: structure-activity relationship. Mol Pharmacol 51: 535–541

Behl C, Widmann M, Trapp T, Holsboer F (1995) 17-β estradiol protects neurons from oxidative stress-induced cell death in vitro. Biochem Biophys Res Com 216: 473–482

Bennett MR (2000) The concept of long term potentiation of transmission at synapses. Prog Neurobiol 60: 109–37

Beral V, Bull D, Doll R, Key T, Peto R, Reeves G, Calle EE, Heath CW, Coates RJ, Liff JM, Franceschi S, Talamini R, Chantarakul N, Koetsawang S, Rachawat D, Morabia A, Schuman L, Stewart W, Szklo M, Bain C, Schofield F, Siskind V, Band P, Coldman AJ, Gallagher RP et al (1997) Breast cancer and hormone replacement therapy – collaborative reanalysis of data from 51 epidemiological studies of 52,705 women with breast cancer and 108,411 women without breast cancer. Lancet 350: 1047–1059

Bernard C (1865) Introduction a la medecine experimentale. Bailliere, Paris

Bethea CL (1993) Colocalization of progestin receptors with serotonin in raphe neurons of macaque. Neuroendocrinology 57: 1–6

Bethea CL, Gundlah C, Mirkes SJ (2000) Ovarian steroid action in the serotonin neural system of macaques. Novartis Foundation Symposium 230, 112–130. John Wiley & Sons, Ltd

Bethea CL, Mirkes SJ, Shively CA, Adams MR (2000) Steroid regulation of tryptophan hydroxylase protein in the dorsal raphe of macaques. Biol Psych 47: 562–76

Beyer C (1999) Estrogen and the developing mammalian brain. Anat Embryol 199: 379–90

Biebl M, Cooper CM, Winkler J, Kuhn HG (2000) Analysis of neurogenesis and programmed cell death reveals a self-renewing capacity in the adult rat brain. Neurosci Lett 291:17–20

Biegon A, Bercovitz H, Samuel D (1980) Serotonin receptor concentration during the estrous cycle of the rat. Brain Res 187: 221–5

Biegon A, Reches A, Snyder L, McEwen BS (1983) Serotonergic and noradrenergic receptors in the rat brain: modulation by chronic exposure to ovarian hormones. Life Sci 32: 2015–21

Biesalski HK, Frank J (1995) Antioxidants in nutrition and their importance for the anti-/pro-oxidative balance in the immune system. Immun Infekt 23: 166–173

Binko J, Majewski H (1998) 17beta-Estradiol reduces vasoconstriction in endothelium-denuded rat aortas through inducible NOS. Am J Physiol 274: H853–9

Bishop J, Simpkins JW (1994) Estradiol treatment increases viability of glioma and neuroblastoma cells in vitro. Mol Cell Neurosci 5: 303–8

Bishop J, Simpkins JW (1995) Estradiol enhances brain glucose uptake in ovariectomized rats. Brain Res Bull 36: 315–20

Blagosklonny MV (1999) A node between proliferation, apoptosis, and growth arrest. Bioessays 21: 704–9

Blanchet PJ, Fang J, Hyland K, Arnold LA, Mouradian MM, Chase TN (1999) Short-term effects of high-dose 17beta-estradiol in postmenopausal PD patients – A crossover study. Neurology 53: 91–95

Blaustein JD, Olster DH (1989) Gonadal steroid hormone receptors and social behaviors. In: Balthazart J (ed) Advances in comparative Environmental Physiology, Vol 3. Springer, Berlin, pp 31–104

Bliss TV, Lomo T (1973) Long-lasting potentiation of synaptic transmission in the dentate area of the anaesthetized rabbit following stimulation of the perforant path. J Physiol 232: 331–56

Blum-Degen D, Haas M, Pohli S, Harth R, Romer W, Oettel M, Riederer P, Gotz ME (1998)

Scavestrogens protect IMR 32 cells from oxidative stress-induced cell death. Toxicol Appl Pharm 152: 49–55

Brace M, McCauley E (1997) Oestrogens and psychological well-being. Ann Med 29: 283–290

Braun S, Liebetrau W, Berning B, Behl C (2000) Vulnerability of clonal hippocampal cells to oxidative stress is increased by dexamethasone and associated with suppression of NF-κB. Neurosci Lett 295: 101–104

Bravo L (1998) Polyphenols: chemistry, dietary sources, metabolism, and nutritional significance. Nutr Rev 56: 317–333

Breckwoldt M, Karck U (2000) Tamoxifen for breast cancer prevention. Exp Clin Endocr Diab 108: 243-246

Breedlove SM (1994) Sexual differentiation of the human nervous system. Annu Rev Psychol 45: 389–418

Breitner JCS (1996) The role of antiinflammatory drugs in the prevention and treatment of Alzheimer's disease. Annu Rev Med 47: 401–411

Brent R (1999) Functional genomics: learning to think about gene expression data. Curr Biol 9: R338–41

Brown PH, Lippman SM (2000) Chemoprevention of breast cancer. Breast Cancer Res Treat 62: 1–17

Brzozowski AM, Pike ACW, Dauter Z, Hubbard RE, Bonn T, Engstrom O, Ohman L, Greene GL, Gustafsson JA, Carlquist M (1997) Molecular basis of agonism and antagonism in the oestrogen receptor. Nature 389 : 753–758

Buffet NC, Djakoure C, Maitre SC, Bouchard P (1998) Regulation of the human menstrual cycle. Front Neuroendocrinol 19: 151–186

Burakov D, Wong CW, Rachez C, Cheskis BJ, Freedman LP (2000) Functional interactions between the estrogen receptor and DRIP205, a subunit of the heteromeric DRIP coactivator complex. J Biol Chem 275: 20928–34

Buratowski S (2000) Snapshots of RNA polymerase II transcription initiation. Curr Opin Cell Biol 12: 320–5

Burton GW, Joyce A, Ingold KU (1983) Is vitamin E the only lipid-soluble, chain-breaking antioxidant in human blood plasma and erythrocyte membranes? Arch Biochem Biophys 221: 281–290

Butenandt A (1929) Untersuchungen über das weibliche Sexualhormon. Deut Med Wochenschr 52: 2171–2173

Cacabelos R, Alvarez A, Lombardi A, Fernandez-Novoa L, Corzo L, Perez P, Laredo M, Pichel V, Hernandez A, Varela M, Figueroa J, Prous J, Windisch M, Vigo C (2000) Pharmacological treatment of Alzheimer disease: From psychotropic drugs and cholinesterase inhibitors to pharmacogenomics drugs. Today 36: 415–499

Callier S, Morissette M, Grandbois M, Di Paolo T (2000) Stereospecific prevention by 17beta-estradiol of MPTP-induced dopamine depletion in mice. Synapse 37: 245–51

Cameron HA, Woolley CS, McEwen BS, Gould E (1993) Differentiation of newly born neurons and glia in the dentate gyrus of the adult rat. Neuroscience 56: 337–44

Campagnoli C, Biglia N, Cantamessa C, Lesca L, Sismondi P (1999) HRT and breast cancer risk: a clue for interpreting the available data. Maturitas 33: 185–190

Campard PK, Crochemore C, Rene F, Monnier D, Koch B, Loeffler JP (1997) PACAP type I receptor activation promotes cerebellar neuron survival through the cAMP/PKA signaling pathway. DNA Cell Biol 16: 323–33

Campbell S, Whitehead M (1977) Oestrogen therapy and the menopausal syndrome. Clin Obstet Gynaecol 4: 31–47

Canada AT, Gianella E, Nguyen TD, Mason RP (1990) The production of reactive oxygen species by dietary flavonols. Free Radical Bio Med 9: 441–449

Cannon WB (1929) Bodily changes in pain, hunger, fear and rage. Appleton, New York

Carlson LE, Sherwin BB (1998) Steroid hormones, memory and mood in a healthy elderly population. Psychoneuroendocrinology 23: 583–603
Carlson LE, Sherwin BB (2000) Higher levels of plasma estradiol and testosterone in healthy elderly men compared with age-matched women may protect aspects of explicit memory. Menopause 7: 168–77
Casper RC (1998) Women's Health: Hormones, Emotions, and Behavior. Cambridge University Press
Cassarino DS, Bennett JP (1999) An evaluation of the role of mitochondria in neurodegenerative diseases: mitochondrial mutations and oxidative pathology, protective nuclear responses, and cell death in neurodegeneration. Brain Res Rev 29: 1–25
Castner SA, Xiao L, Becker JB (1993) Sex differences in striatal dopamine: in vivo microdialysis and behavioral studies. Brain Res 610: 127–34
Cattaneo E, McKay R (1990) Proliferation and differentiation of neuronal stem cells regulated by nerve growth factor. Nature 347: 762–5
Caulin-Glaser T, Watson CA, Pardi R, Bender JR (1996) Effects of 17beta-estradiol on cytokine-induced endothelial cell adhesion molecule expression. J Clin Invest 98: 36–42
Celis JE, Kruhoffer M, Gromova I, Frederiksen C, Ostergaard M, Thykjaer T, Gromov P, Yu J, Palsdottir H, Magnusson N, Orntoft TF (2000) Gene expression profiling: monitoring transcription and translation products using DNA microarrays and proteomics. FEBS Lett 480: 2–16
Chabbert Buffet N, Djakoure C, Maitre SC, Bouchard P (1998) Regulation of the human menstrual cycle. Front Neuroendocrin 19: 151–86
Chee M, Yang R, Hubbell E, Berno A, Huang XC, Stern D, Winkler J, Lockhart DJ, Morris MS, Fodor SP (1996) Accessing genetic information with high-density DNA arrays. Science 274: 610–4
Chen Z, Yuhanna IS, Galcheva-Gargova Z, Karas RH, Mendelsohn ME, Shaul PWl (1999) Estrogen receptor alpha mediates the nongenomic activation of endothelial nitric oxide synthase by estrogen. J Clin Invest 103: 401–6
Chiechi LM, Secreto G (2000) Factors of risk for breast cancer influencing postmenopausal long-term hormone replacement therapy. Tumori 86: 12–16
Chisolm GM, Steinberg D (2000) The oxidative modification hypothesis of atherogenesis: An overview. Free Radic Biol Med 28: 1815–1826
Chomicka LK (1986) Effect of oestradiol on the responses of regional brain serotonin to stress in the ovariectomized rat. J Neural Transm 67: 267–73
Christ M, Meyer C, Sippel Kl, Wehling M (1995) Rapid aldosterone signaling in vascular smooth muscle cells: involvement of phospholipase C, diacylglycerol and protein kinase C alpha. Biochem Biophys Res Commun 213: 123–9
Christen S, Woodall AA, Shigenaga MK, Southwell-Keely PT, Duncan MW, Ames BN (1997) γ-Tocopherol traps mutagenic electrophiles such as NO_X and complements α-tocopherol: Physiological implications. Proc Natl Acad Sci USA 94: 3217–3222
Chu S, Fuller PJ (1997) Identification of a splice variant of the rat estrogen receptor beta gene. Mol Cell Endocrinol 132: 195–9
Chu S, Mamers P, Burger HG, Fuller PJ (2000) Estrogen receptor isoform gene expression in ovarian stromal and epithelial tumors. J Clin Endocrinol Metab 85: 1200–5
Ciocca DR, Fanelli MA (1997) Estrogen receptors and cell proliferation in breast cancer. Trends Endocrin Met 8: 313–321
Ciocca DR, Roig LMV (1995) Estrogen receptors in human nontarget tissues – biological and clinical implications. Endocrin Rev 16: 35–62
Ciolino HP, Daschner PJ, Yeh GC (1998) Resveratrol inhibits transcription of CYP1A1 in vitro by preventing activation of the aryl hydrocarbon receptor. Cancer Res 58: 5707–5712
Clark JH, Peck EJ Jr (1979) Female sex steroids: receptors and function. Monogr Endocrinol 14: I–XII, 1–245

Clement MV, Hirpara JL, Chawdhury SH, Pervaiz S (1998) Chemopreventive agent resveratrol, a natural product derived from grapes, triggers CD95 signaling-dependent apoptosis in human tumor cells. Blood 92: 996–1002

Clemett D, Spencer CM (2000) Raloxifene – A review of its use in postmenopausal osteoporosis. Drugs 60: 379–411

Cohen RZ, Seeman MV, Gotowiec A, Kopala L (1999) Earlier puberty as a predictor of later onset of schizophrenia in women. Am J Psychiatry 156: 1059–1064

Colditz GA (1998) Relationship between estrogen levels, use of hormone replacement therapy, and breast cancer. J Natl Cancer I 90: 814–823

Collingwood TN, Urnov FD, Wolffe AP (1999) Nuclear receptors: coactivators, corepressors and chromatin remodeling in the control of transcription. J Mol Endocrinol 23: 255–75

Collins FS, Patrinos A, Jordan E, Chakravarti A, Gesteland R, Walters L (1998) New goals for the U.S. Human Genome Project: 1998–2003. Science 282: 682–9

Cone RI, Davis GA, Goy RW (1981) Effects of ovarian steroids on serotonin metabolism within grossly dissected and microdissected brain regions of the ovariectomized rat. Brain Res Bull 7: 639–44

Cookson MR, Shaw PJ (1999) Oxidative stress and motor neurone disease. Brain Pathol 9: 165–186

Cordoba Montoya DA, Carrer HF (1997) Estrogen facilitates induction of long term potentiation in the hippocampus of awake rats. Brain Res 778: 430–8

Corner GW (1964) The early history of the oestrogenic hormones. The Sir Henry Dale Lecture for 1964

Couse JF, Korach KS (1999) Estrogen receptor null mice: what have we learned and where will they lead us? Endocr Rev 20: 358–417

Couse JF, Korach KS (1999) Estrogen receptor null mice: what have we learned and where will they lead us? Endocr Rev 20: 358–417

Coyle JT, Puttfarcken P (1993) Oxidative stress, glutamate, and neurodegenerative disorders. Science 262: 689–695

Craig CG, Tropepe V, Morshead CM, Reynolds BA, Weiss S, van der Kooy D (1996) In vivo growth factor expansion of endogenous subependymal neural precursor cell populations in the adult mouse brain. J Neurosci 16: 2649–58

Craik FIM, Jennings JM (1992) Human memory. In: Craik FIM, Salthouse TA (eds) The handbook of Aging and Cognition Earlbaum. Hillsdale, NJ, pp 51–110

Cramer SC, Bastings EP (2000) Mapping clinically relevant plasticity after stroke. Neuropharmacol 39: 842–51

Cross AR, Jones OTG (1991) Enzymatic mechanisms of superoxide production. Biochim Biophys Acta 1057: 281–298

Cross TG, Scheel-Toellner D, Henriquez NV, Deacon E, Salmon M, Lord JM (2000) Serine/threonine protein kinases and apoptosis. Exp Cell Res 256: 34–41

Cummings JL, Back C (1998) The cholinergic hypothesis of neuropsychiatric symptoms in Alzheimer's disease. Am J Geriatr Psych 6 [Suppl 1]: S64–78

Curtis SH, Korach KS (2000) Steroid receptor knockout models: phenotypes and responses illustrate interactions between receptor signaling pathways in vivo. Adv Pharmacol (New York) 47: 357–80

Cynshi O, Kawabe Y, Suzuki T, Takashima Y, Kaise H, Nakamura M, Ohba Y, Kato Y, Tamura K, Hayasaka A et al (1998) Antiatherogenic effects of the antioxidant BO-653 in three different animal models. Proc Natl Acad Sci USA 95: 10123–10128

Cyr M, Calon F, Morissette M, Grandbois M, Callier S, Di Paolo T (2000) Drugs with estrogen-like potency and brain activity: Potential therapeutic application for the CNS Current Pharmaceutical Design 6: 1287–1312

Daniel JM, Fader AJ, Spencer AL, Dohanich GP (1997) Estrogen enhances performance of female rats during acquisition of a radial arm maze. Horm Behav 32: 217–25

Das NP (1971) Studies on flavonoid metabolism. Absorption and metabolism of (+)-catechin in man. Biochem Pharmacol 20: 3435–3445

Dasilva SL, Burbach JPH (1995) The Nuclear Hormone-Receptor Family In The Brain – Classics And Orphans. Trends Neurosci 18: 542–548

De Vries JH, Hollman PC, Meyboom S, Buysman MN, Zock PL, van Staveren, WA, Katan MB (1998) Plasma concentrations and urinary excretion of the antioxidant flavonols quercetin and kaempferol as biomarkers for dietary intake. Am J Clin Nutr 68: 60–65

De Vries JH, Janssen K, Hollman PC, van Staveren WA, Katan MB (1997) Consumption of quercetin and kaempferol in free-living subjects eating a variety of diets. Cancer Lett 114: 141–144

Delmas PD, Bjarnason NH, Mitlak BH, Ravoux AC, Shah AS, Huster WJ, Draper M, Christiansen C (1997) Effects of raloxifene on bone mineral density, serum cholesterol concentrations, and uterine endometrium in postmenopausal women. N Engl J Med 337: 1641–1647

Deng G, Pike CJ, Cotman CW (1996) Alzheimer-associated presenilin-2 confers increased sensitivity to apoptosis in PC12 cells. FEBS Lett 397: 50–4

Derkinderen P, Enslen H, Girault JA (1999) The ERK/MAP-kinases cascade in the nervous system. Neuroreport 10: R24–34

Desagher S, Martinou JC (2000) Mitochondria as the central control point of apoptosis. Trends Cell Biol 10: 369–377

DeSombre ER, Greene GL, King WJ, Jensen EV (1984) Estrogen receptors, antibodies and hormone dependent cancer. Progr Clin Biol Res 142: 1–21

Deveraux QL, Reed JC (1999) IAP family proteins-suppressors of apoptosis. Gen Dev 13: 239–52

Dhingra K (1999) Antiestrogens – tamoxifen, SERMs and beyond. Invest New Drug 17: 285–311

Di Paolo T, Rouillard C, Bedard P (1985) 17beta-Estradiol at a physiological dose acutely increases dopamine turnover in rat brain. Eur J Pharmacol 117: 197–203

Diamond MC (1991) Hormonal effects on the development or cerebral lateralization. Psychoneuroendocrinology 16: 121–9

Dickancaite E, Nemeikaite A, Kalvelyte A, Cenas N (1998) Prooxidant character of flavonoid cytotoxicity: structure-activity relationships. Biochem Mol Biol Int 45: 923–930

Dickinson SL, Curzon G (1986) 5-Hydroxytryptamine-mediated behaviour in male and female rats. Neuropharmacology 25: 771–6

Dirnagl U, Iadecola C, Moskowitz MA (1999) Pathobiology of ischaemic stroke: an integrated view. Trends Neurosci 22: 391–397

D'Mello SR, Galli C, Ciotti T, Calissano P (1993) Induction of apoptosis in cerebellar granule neurons by low potassium: inhibition of death by insulin-like growth factor I and cAMP. Proc Natl Acad Sci USA 90: 10989–10993

Dodel RC, Du YS, Bales KR, Gao F, Paul SM (1999) Sodium salicylate and 17beta-estradiol attenuate nuclear transcription factor NF-kappa B translocation in cultured rat astroglial cultures following exposure to amyloid A beta(1–40) and lipopolysaccharides. J Neurochem 73: 1453–1460

Dong L, Wang W, Wang F, Stoner M, Reed JC, Harigai M, Samudio I, Kladde MP, Vyhlidal C, Safe S (1999) Mechanisms of transcriptional activation of BCL-2 gene expression by 17beta-estradiol in breast cancer cells. J Biol Chem 274: 32099–107

Drake EB, Henderson VW, Stanczyk FZ, McCleary CA, Brown WS, Smith CA, Rizzo AA, Murdock GA, Buckwalter JG (2000) Associations between circulating sex steroid hormones and cognition in normal elderly women. Neurology 54: 599–603

Drewes G, Lichtenberg-Kraag B, Doring F, Mandelkow EM, Biernat J, Goris J, Doree M, Mandelkow E (1992) Mitogen activated protein (MAP) kinase transforms tau protein into an Alzheimer-like state. EMBO J 11: 2131–8

Driscoll MD, Sathya G, Muyan M, Klinge CM, Hilf R, Bambara RA (1998) Sequence require ments for estrogen receptor binding to estrogen response elements. J Biol Chem 273: 29321–30

Dubal DB, Kashon ML, Pettigrew LC, Ren JM, Finklestein SP, Rau SW, Wise PM (1998) Estradiol protects against ischemic injury. J Cereb Blood Flow Metab 18: 1253–1258

Dubal DB, Shughrue PJ, Wilson ME, Merchenthaler I, Wise PM (1999) Estradiol modulates bcl-2 in cerebral ischemia: a potential role for estrogen receptors. J Neurosci 19: 6385–93

Dudek H, Datta SR, Franke TF, Birnbaum MJ, Yao RJ, Cooper GM, Segal RA, Kaplan DR, Greenberg ME (1997) Regulation of neuronal survival by the serine-threonine protein kinase akt. Science 275: 661–665

Dumont A, Hehner SP, Schmitz ML, Gustafsson JA, Liden J, Okret S, van der Saag PT, Wissink S, van der Burg B, Herrlich P, Haegeman G, De Bosscher K, Fiers W (1998) Cross-talk between steroids and NF-kappa B: what language? Trends Biochem Sci 23: 233–5

Dunning AM, Healey CS, Pharoah PDP, Teare MD, Ponder BAJ, Easton DF (1999) A systematic review of genetic polymorphisms and breast cancer risk. Cancer Epidem Biomar 8: 843–854

Ebadi M, Srinivasan SK, Baxi MD (1996) Oxidative stress and antioxidant therapy in Parkinson's disease. Progress in Neurobiology 48: 1–19

Eikelenboom P, Rozemuller AJM, Hoozemans JJM, Veerhuis R, van Gool WA (2000) Neuroinflammation and Alzheimer disease: Clinical and therapeutic implications. Alz Dis Ass Dis 14 [Suppl 1]: S54–S61

El Kossi MMH, Zakhary MM (2000) Oxidative stress in the context of acute cerebrovascular stroke. Stroke 31: 1889–1892

Ellis RJ, Hartl FU (1999) Principles of protein folding in the cellular environment. Curr Opin Struc Biol 9: 102–10

Enmark E, Pelto-Huikko M, Grandien K, Lagercrantz S, Lagercrantz J, Fried G, Nordenskjold M, Gustafsson JA (1997) Human estrogen receptor-gene structure, chromosomal localization, and expression pattern. J Clin Endocrinol Metab 82: 4258–65

Eriksson PS, Perfilieva E, Bjork-Eriksson T, Alborn AM, Nordborg C, Peterson DA, Gage FH (1998) Neurogenesis in the adult human hippocampus. Nat Med 4: 1313–7

Esterbauer H, Schaur RJ, Zollner H (1991) Chemistry and biochemistry of 4-hydroxynonenal, malonaldehyde and related aldehydes. Free Radical Bio Med 11: 81–128

Evans DA, Funkenstein HH, Albert MS, Scherr PA, Cook NR, Chown MJ, Hebert LE, Hennekens CH, Taylor JO (1989) Prevalence of Alzheimer's disease in a community population of older persons. J Am Med Ass 262: 2551–2556

Evans RM (1988) The steroid and thyroid hormone receptor superfamily. Science 240: 889–95

Fahn S (1999) Parkinson disease, the effect of levodopa, and the ELLDOPA trial. Arch Neurol 56: 529–535

Farr SA, Banks WA, Morley JE (2000) Estradiol potentiates acetylcholine and glutamate-mediated post-trial memory processing in the hippocampus. Brain Res 864: 263–9

Fauconneau B, Waffo-Teguo P, Huguet F, Barrier L, Decendit A, Merillon JM (1997) Comparative study of radical scavenger and antioxidant properties of phenolic compounds from *vitis vinifera* cell cultures using *in vitro* tests. Life Sci 61: 2103–2110

Fillit H, Weinreb H, Cholst I, Luine V, McEwen B, Amador R, Zabriskie J (1986) Observations in a preliminary open trial of estradiol therapy for senile dementia-Alzheimer's type. Psychoneuroendocrinology 11: 337–45

Finch CE, Cohen DM (1997) Aging metabolism, and Alzheimer disease – review and hypotheses. Exp Neurol 143: 82–102

Fink BE, Mortensen DS, Stauffer SR, Aron ZD, Katzenellenbogen JA (1999) Novel structural templates for estrogen-receptor ligands and prospects for combinatorial synthesis of estrogens. Chem Biol 6: 205–19

Fink G, Sumner BE, Rosie R, Grace O, Quinn JP (1996) Estrogen control of central neurotransmission: effect on mood, mental state, and memory. Cell Mol Neurobiol 16: 325–44

Finkbeiner S (2000) Calcium regulation of the brain-derived neurotrophic factor gene. Cell Mol Life Sci 57: 394–401

Finkbeiner S (2000) CREB couples neurotrophin signals to survival messages. Neuron 25: 11–14

Finucane FF, Madans JH, Bush TL, Wolf PH, Kleinman JC (1993) Decreased risk of stroke among postmenopausal hormone users. Results from a national cohort. Arch Intern Med 153: 73–9

Fiorelli G, Gori F, Frediani U, Franceschelli F, Tanini A, Tosti-Guerra C, Benvenuti S, Gennari L, Becherini L, Brandi ML Membrane binding sites and non-genomic effects of estrogen in cultured human pre-osteoclastic cells. J Steroid Biochem & Mol Biol 59: 233–40 (1996)

Flood JF, Morley JE, Roberts E (1992) Memory-enhancing effects in male mice of pregnenolone and steroids metabolically derived from it. Proc Natl Acad Sci USA 89: 1567–71

Floor E (2000) Iron as a vulnerability factor in nigrostriatal degeneration in aging and Parkinson's disease. Cell Mol Biol 46: 709–720

Flynn BL, Ranno AE (1999) Pharmacologic management of Alzheimer disease part II: Antioxidants, antihypertensives, and ergoloid derivatives. Ann Pharmacotherapy 33: 188–197

Foley P, Riederer P (2000) Influence of neurotoxins and oxidative stress on the onset and progression of Parkinson's disease. J Neurol 247 [Suppl 2]: II82–94

Franke TF, Yang SI, Chan TO, Datta K, Kazlauskas A, Morrison DK, Kaplan DR, Tsichlis PN (1995) The protein kinase encoded by the akt proto-oncogene is a target of the pdgf-activated phosphatidylinositol 3-kinase. Cell 81: 727–736

Frankfurt M, Gould E, Woolley CS, McEwen BS (1990) Gonadal steroids modify dendritic spine density in ventromedial hypothalamic neurons: a Golgi study in the adult rat. Neuroendocrinology 51: 530–5

Freedman LP (1999) Strategies for transcriptional activation by steroid/nuclear receptors. J Cell Biochem [Suppl 32-33]: 103–9

Friedhoff P, von Bergen M, Mandelkow EM, Mandelkow E (2000) Structure of tau protein and assembly into paired helical filaments. Biochim Biophys Acta – Molecular Basis of Disease 1502: 122–132

Friedman WJ, Greene LA (1999) Neurotrophin signaling via Trks and p75. Exp Cell Res 253: 131–42

Frohlich J, Ogawa S, Morgan M, Burton L, Pfaff D (1999) Hormones, genes and the structure of sexual arousal. Behav Brain Res 105: 5–27

Frye CA, Duncan JE (1994) Progesterone metabolites, effective at the GABAA receptor complex, attenuate pain sensitivity in rats. Brain Res 643: 194–203

Fuhrmann U, Parczyk K, Klotzbucher M, Klocker H, Cato AC (1998) Recent developments in molecular action of antihormones. J Mol Med 76: 512–24

Gage FH (1994) Neuronal stem cells: their characterization and utilization. Neurobiol Aging 15 [Suppl 2]: S191

Gage FH (1998) Cell therapy. Nature 392 [6679 Suppl S]: 18–24

Gage FH (2000) Mammalian neural stem cells. Science 287: 1433–1438

Galea LA, Kavaliers M, Ossenkopp KP (1996) Sexually dimorphic spatial learning in meadow voles Microtus pennsylvanicus and deer mice Peromyscus maniculatus. J Exp Biol 199 (Pt 1): 195–200

Games D, Adams D, Alessandrini R, Barbour R, Berthelette P, Blackwell C, Carr T, Clemens J, Donaldson T, Gillespie F et al (1995) Alzheimer-type neuropathology in transgenic mice overexpressing V717F beta-amyloid precursor protein. Nature 373: 523–7

Gandy S (1999) Neurohormonal signaling pathways and the regulation of Alzheimer beta-amyloid precursor metabolism. Trends Endocrin Met 10: 273–279

Garcia-Segura LM, Cardona-Gomez P, Naftolin F, Chowen JA (1998) Estradiol upregulates BCL-2 expression in adult brain neurons. Neuroreport 9: 593–7

Garcia-Segura LM, Naftolin F, Hutchison JB, Azcoitia I, Chowen JA (1999) Role of astroglia in estrogen regulation of synaptic plasticity and brain repair. J Neurobiol 40: 574–84

Garnier M, lorenzo D, Albertini A, Maggi A (1997) Identification of estrogen-responsive genes in neruoblastoma SK-ER3 cells. J Neurosci 17: 4591–4599

Gate L, Paul J, Ba GN, Tew KD, Tapiero H (1999) Oxidative stress induced in pathologies: the role of antioxidants. Biomed Pharmacother 53: 169–180

Gee KW, Bolger MB, Brinton RE, Coirini H, McEwen BS (1988) Steroid modulation of the chloride ionophore in rat brain: structure-activity requirements, regional dependence and mechanism of action. J Pharmacol Exp Ther 246: 803–12

Genazzani AR, Gastaldi M, Bidzinska B, Mercuri N, Genazzani AD, Nappi RE, Segre A, Petraglia F (1992) The brain as a target organ of gonadal steroids. Psychoneuroendocrinology 17: 385–90

Ghosh A, Greenberg ME (1995) Calcium signaling in neurons – molecular mechanisms and cellular consequences. Science 268: 239–247

Giasson BI, Duda JE, Murray IV, Chen Q, Souza JM, Hurtig HI, Ischiropoulos H, Trojanowski JQ, Lee VM (2000) Oxidative damage linked to neurodegeneration by selective alpha-synuclein nitration in synucleinopathy lesions. Science 290: 985–989

Gibbs RB (2000) Oestrogen and the cholinergic hypothesis: implications for oestrogen replacement therapy in postmenopausal women, Novartis Foundation Symposium 230: 94-107, John Wiley & Sons, Ltd.

Gibbs RB, Aggarwal P (1998) Estrogen and basal forebrain cholinergic neurons: implications for brain aging and Alzheimer's disease-related cognitive decline. Horm Behav 34: 98–111

Giguere V (1999) Orphan nuclear receptors: From gene to function. Endocr Rev 20: 689–725

Giguere V, Tremblay A, Tremblay GB (1998) Estrogen receptor beta: re-evaluation of estrogen and antiestrogen signaling. Steroids 63: 335–9

Giguere V, Yang N, Segui P, Evans RM (1988) Identification of a new class of steroid hormone receptors. Nature 331: 91–4

Gilligan DM, Quyyumi AA, Cannon RO 3rd (1994) Effects of physiological levels of estrogen on coronary vasomotor function in postmenopausal women. Circulation 89: 2545–51

Gillum LA, Mamidipudi SK, Johnston SC (2000) Ischemic stroke risk with oral contraceptives – A meta-analysis. Jama 284: 72–78

Glenner GG, Wong CW (1984) Alzheimer's disease: initial report of the purification and characterization of a novel cerebrovascular amyloid protein. Biochem Biophys Res Com 120: 885–890

Goodenough S, Engert S, Behl C (2000) Testosterone stimulates rapid secretory amyloid precursor protein release from rat hypothalamic cells via the activation of the mitogen-activated protein kinase pathway. Neurosci Lett 296: 49–52

Goodman RH, Smolik S (2000) CBP/p300 in cell growth, transformation, and development. Gen Dev 14: 1553–1577

Goodman Y, Bruce AJ, Cheng B, Mattson MP (1996) Estrogens attenuate and corticosterone exacerbates excitotoxicity, oxidative injury, and amyloid beta-peptide toxicity in hippocampal neurons. J Neurochem 66: 1836–44

Gould E, Beylin A, Tanapat P, Reeves A, Shors TJ (1999a) Learning enhances adult neurogenesis in the hippocampal formation. Nat Neurosci 2: 260–5

Gould E, McEwen BS, Tanapat P, Galea LA, Fuchs E (1997) Neurogenesis in the dentate gyrus of the adult tree shrew is regulated by psychosocial stress and NMDA receptor activation. J Neurosci 17: 2492–8

Gould E, Reeves AJ, Fallah M, Tanapat P, Gross CG, Fuchs E (1999b) Hippocampal neurogenesis in adult Old World primates. Proc Natl Acad Sci USA 96: 5263–7

Gould E, Tanapat P, McEwen BS, Flugge G, Fuchs E (1998) Proliferation of granule cell precursors in the dentate gyrus of adult monkeys is diminished by stress. Proc Natl Acad Sci USA 95: 3168–71

Gouras GK, Xu HX, Gross RS, Greenfield JP, Hai B, Wang R, Greengard P (2000) Testosterone reduces neuronal secretion of Alzheimer's beta-amyloid peptides. Proc Natl Acad Sci USA 97: 1202–1205

Grady D, Rubin SM, Petitti DB, Fox CS, Black D, Ettinger B, Ernster VL, Cummings SR (1992) Hormone therapy to prevent disease and prolong life in postmenopausal women. Ann Intern Med 117: 1016–37

Grandbois M, Morissette M, Callier S, Di Paolo T (2000) Ovarian steroids and raloxifene prevent MPTP-induced dopamine depletion in mice. Neuroreport 11: 343–6

Grazzini E, Guillon G, Mouillac B, Zingg HH (1998) Inhibition of oxytocin receptor function by direct binding of progesterone. Nature 392: 509–12

Green PS, Bishop J, Simpkins JW (1997) 17 alpha-estradiol exerts neuroprotective effects on SK-N-SH cells. J Neurosci 17: 511–5

Green PS, Gordon K, Simpkins JW (1997) Phenolic A ring requirement for the neuroprotective effects of steroids. J Steroid Biochem 63: 229–235

Green PS, Simpkins JW (2000) Neuroprotective effects of estrogens: potential mechanisms of action. Intern J Develop Neurosci 18: 347–358

Green S, Chambon P (1988) Nuclear receptors enhance our understanding of transcription regulation. Trends Genet 4: 309–14

Green S, Walter P, Kumar V, Krust A, Bornert JM, Argos P, Chambon P (1986) Human oestrogen receptor cDNA: sequence, expression and homology to v-erb-A. Nature 320: 134–9

Greene GL, Closs LE, Fleming H, DeSombre ER, Jensen EV (1977) Antibodies to estrogen receptor: immunochemical similarity of estrophilin from various mammalian species. Proc Natl Acad Sci USA 74: 3681–5

Greene GL, Sobel NB, King WJ, Jensen EV (1984) Immunochemical studies of estrogen receptors. J Steroid Biochem 20: 51–6

Gridley KE, Green PS, Simpkins JW (1997) Low concentrations of estradiol reduce beta-amyloid (25–35)-induced toxicity, lipid peroxidation and glucose utilization in human SK-N-SH neuroblastoma cells. Brain Res 778:158–65

Gridley KE, Green PS, Simpkins JW (1998) A novel, synergistic interaction between 17beta-estradiol and glutathione in the protection of neurons against beta-amyloid 25-35-induced toxicity in vitro. Mol Pharmacol 54: 874–880

Grilli M, Pizzi M, Memo M, Spano P (1996) Neuroprotection by aspirin and sodium salicylate though blockade of NF-κB activation. Science 274: 1383–1385

Grodstein F, Stampfer MJ, Colditz GA, Willett WC, Manson JE, Joffe M, Rosner B, Fuchs C, Hankinson SE, Hunter DJ, Hennekens CH, Speizer FE (1997) Postmenopausal hormone therapy and mortality. N Engl J Med 336: 1769–75

Grodstein F, Stampfer MJ, Manson JE, Colditz GA, Willett WC, Rosner B, Speizer FE, Hennekens CH (1996) Postmenopausal estrogen and progestin use and the risk of cardiovascular disease. N Engl J Med 335: 453–61

Groner B, Hynes NE, Rahmsdorf U, Ponta H (1983) Transcription initiation of transfected mouse mammary tumor virus LTR DNA is regulated byglucocorticoid hormones. Nucl Acids Res 11: 4713–25

Gu G, Rojo AA, Zee MC, Yu J, Simerly RB (1996) Hormonal regulation of CREB phosphorylation in the anteroventral periventricular nucleus. J Neurosci 16: 3035–44

Gu Q, Moss RL (1996) 17beta-Estradiol potentiates kainate-induced currents via activation of the cAMP cascade. J Neurosci 16: 3620–9

Gudino-Cabrera G, Nieto-Sampedro M (1999) Estrogen receptor immunoreactivity in Schwann-like brain macroglia. J Neurobiol 40: 458–70

Gugler R, Leschik M, Dengler HJ (1975) Disposition of quercetin in man after single oral and intravenous doses. Eur J Clin Pharmacol 9: 229–234

Gur RC, Gur RE, Obrist WD, Hungerbuhler JP, Younkin D, Rosen AD, Skolnick BE Reivich M (1982) Sex and handedness differences in cerebral blood flow during rest and cognitive activity. Science 217: 659–61

Gur RC, Mozley LH, Mozley PD, Resnick SM, Karp JS, Alavi A, Arnold SE, Gur RE (1995) Sex differences in regional cerebral glucose metabolism during a resting state. Science 267: 528–31

Gustafsson JA (1999) Estrogen receptor beta-a new dimension in estrogen mechanism of action. J Endocrinol 163: 379–83

Gustafsson JA (2000) An update on estrogen receptors. Semin Perinatol 24: 66–9

Haass C (1997) Presenilins: genes for life and death. Neuron 18: 687–90

Haass C, Mandelkow E (1999) Proteolysis by presenilins and the renaissance of tau. Trends Cell Biol 9: 241–4

Haass C, Selkoe DJ (1993) Cellular processing of ß-amyloid precursor protein and the genesis of amylod β-peptide. Cell 75: 1039–1042

Hacia JG, Brody LC, Chee MS, Fodor SP, Collins FS (1996) Detection of heterozygous mutations in BRCA1 using high density oligonucleotide arrays and two-colour fluorescence analysis. Nature Genet 14: 441–7

Hackman BW, Galbraith D (1976) Replacement therapy and piperazine oestrone sulphate ('Harmogen') and its effect on memory. Curr Med Res Opin 4: 303–6

Hafner H, Anderheiden W (1997) Epidemiology of schizophrenia. Can J Psychiatry 42: 139–151

Halbreich U, Rojansky N, Palter S, Tworek H, Hissin P, Wang K (1995) Estrogen augments serotonergic activity in postmenopausal women. Biol Psychiatry 37: 434–41

Haleem DJ, Kennett GA, Curzon G (1990) Hippocampal 5-hydroxytryptamine synthesis is greater in female rats than in males and more decreased by the 5-HT1A agonist 8-OH-DPAT. J Neural Transm – Gen 79: 93–101

Hall ED, Pazara KE, Linseman KL (1991) Sex differences in postischemic neuronal necrosis in gerbils. J Cereb Blood Flow Metab 11: 292–8

Halliday G, Robinson SR, Shepherd C, Kril J (2000) Alzheimer's disease and inflammation: a review of cellular and therapeutic mechanisms. Clin Exp Pharmacol Physiol 27: 1–8

Halliday GM, Shepherd CE, McCann H, Reid WGJ, Grayson DA, Broe GA, Kril JJ (2000) Effect of anti-inflammatory medications on neuropathological findings in Alzheimer disease. Arch Neurol 7: 831–836

Halliwell B (2000) The antioxidant paradox. Lancet 355: 1179–1180

Hammerschmidt DE, Knabe AC, Silberstein PT, Lamche HR, Coppo PA (1988) Inhibition of granulocyte function by steroids is not limited to corticoids. Studies with sex steroids. Inflammation 12: 277–84

Hammond CB (1996) Menopause and hormone replacement therapy: an overview. Obstet Gynecol 87 [2 Suppl]: 2S–15S

Hampson AJ, Grimaldi M, Axelrod J, Wink D (1998) Cannabidiol and (–)Δ^9-tetrahydrocannabinol are neuroprotective antioxidants. Proc Natl Acad Sci USA 95: 8268–8273

Hansen MB, Nielsen SE, Berg K (1989) Re-examination and further development of a precise and rapid dye method for measuring cell growth/killing. J Immunol Methods 119: 203–210

Hanson MG Jr, Shen S, Wiemelt AP, McMorris FA, Barres BA (1998) Cyclic AMP elevation is sufficient to promote the survival of spinal motor neurons in vitro. J Neurosci 18: 7361–71

Hanstein B, Liu H, Yancisin MC, Brown M (1999) Functional analysis of a novel estrogen receptor-beta isoform. Mol Endocrinol 13: 129–37

Hardy J (1997) Amyloid, the presenilins and Alzheimer's disease. Trends Neurosci 20: 154–159

Harlan RE (1988) Regulation of neuropeptide gene expression by steroid hormones. Mol Neurobiol 2: 183–200

Harman D (1998) Aging: phenomena and theories. Ann NY Acad Sci 854: 1–7

Harnish DC, Scicchitano MS, Adelman SJ, Lyttle CR, Karathanasis SK (2000) The role of CBP in estrogen receptor cross-talk with nuclear factor-kappa B in HepG2 cells. Endocrinology 141: 3403–3411

Harris ME, Hensley K, Butterfield DA, Leedle RA, Carney JM (1995) Direct evidence of oxidative injury produced by the Alzheimer's β-amyloid peptide (1-40) in cultured hippocampal neurons. Exp Neurol 131: 193–202

Hassan A, Markus HS (2000) Genetics and ischaemic stroke. Brain 123: 1784–1812

Heck S, Lezoualc'h F, Engert S, Behl C (1999) Insulin-like growth factor-1-mediated neuroprotection against oxidative stress is associated with activation of nuclear factor kappa B. J Biol Chem 274: 9828–9835

Hippeli S, Elstner EF (1997) OH-radical-type reactive oxygen species: a short review on the mechanisms of OH-radical- and peroxynitrite toxicity. Zeitschrift für Naturforschung. Section C. Journal of Biosciences 52: 555–563

Heller W (1993) Gender differences in depression: perspectives from neuropsychology. J Affect Disorders 29: 129–43

Henderson VW, Paganini-Hill A, Miller BL, Elble RJ, Reyes PF, Shoupe D, McCleary CA, Klein RA, Hake AM, Farlow MR (2000) Estrogen for Alzheimer's disease in women: randomized, double-blind, placebo-controlled trial. Neurology 54: 295–301

Henderson VW, Paganini-Hill A, Emanuel CK, Dunn ME, Buckwalter JG (1994) Estrogen replacement therapy in older women. Arch Neurol 51: 896–900

Herbison AE (1998) Multimodal influence of estrogen upon Gonadotropin-releasing hormone neurons. Endocr Rev 19: 302–330

Heritage AS, Stumpf WE, Sar M, Grant LD (1980) Brainstem catecholamine neurons are target sites for sex steroid hormones. Science 207: 1377–9

Heyman A, Wilkinson WE, Stafford JA, Helms MJ, Sigmon AH, Weinberg T (1984) Alzheimer's disease: a study of epidemiological aspects. Ann Neurol 15: 335–41

Holliday R (1998) Causes of aging. Ann NY Acad Sci 854: 61–71

Hollman PC, van Trijp JM, Mengelers MJ, de Vries JH, Katan MB (1997) Bioavailability of the dietary antioxidant flavonol quercetin in man. Cancer Lett 114: 139–140

Holsboer F, Barden N (1996) Antidepressants and hypothalamic pituitary adrenocortical regulation. Endocr Rev 17: 187–205

Honjo H, Ogino Y, Naitoh K, Urabe M, Kitawaki J, Yasuda J, Yamamoto T, Ishihara S, Okada H, Yonezawa T et al (1989) In vivo effects by estrone sulfate on the central nervous system-senile dementia (Alzheimer's type). J Steroid Biochem 34: 521–5

Horvath TL, Garciasegura LM, Naftolin F (1997) Control of gonadotropin feedback – the possible role of estrogen-induced hypothalamic synaptic plasticity. Gynecol Endocrinol 11: 139–143

Horwitz KB, Jackson TA, Bain DL, Richer JK, Takimoto GS, Tung L (1996) Nuclear receptor coactivators and corepressors. Mol Endocrinol 10: 1167–77

Hosli E, Hosli L (1999) Cellular localization of estrogen receptors on neurones in various regions of cultured rat CNS: coexistence with cholinergic and galanin receptors. Int J Dev Neurosci 17: 317–30

Howell A, Osborne CK, Morris C, Wakeling AE (2000) ICI 182,780 (Faslodex): development of a novel, "pure" antiestrogen. Cancer 89: 817–25

Hoyer S (1995) Age-related changes in cerebral oxidative metabolism – implications for drug therapy. Drug Aging 6: 210–218

Hsu SY, Hsueh AJ (2000) Discovering new hormones, receptors, and signaling mediators in the genomic era. Mol Endocrinol 14: 594–604

Hunt K, Vessey M, McPherson K (1990) Mortality in a cohort of long-term users of hormone replacement therapy: an updated analysis. Br J Obstet Gynaecol 97: 1080–6

Hurn PD, Macrae IM (2000) Estrogen as a neuroprotectant in stroke. J Cereb Blood Flow Metab 20: 631–652

Hutchison JB (1993) Aromatase: neuromodulator in the control of behavior. J Steroid Biochem Mol Biol 44: 509–20

Iafrati MD, Karas RH, Aronovitz M, Kim S, Sullivan TR Jr, Lubahn DB, O'Donnell TF Jr, Korach

KS, Mendelsohn ME (1997) Estrogen inhibits the vascular injury response in estrogen receptor alpha-deficient mice. Nat Med 3: 545–8

Ignarro LJ (1999) Nitric oxide: a unique endogenous signaling molecule in vascular biology. Bioscience Rep 19: 51–71

Improta-Brears T, Whorton AR, Codazzi F, York JD, Meyer T, McDonnell DP (1999) Estrogen-induced activation of mitogen-activated protein kinase requires mobilization of intracellular calcium. Proc Natl Acad Sci USA 96: 4686–91

Irizarry MC, Soriano F, McNamara M, Page KJ, Schenk D, Games D, Hyman BT (1997) A-beta deposition is associated with neuropil changes, but not with overt neuronal loss in the human amyloid precursor protein v717f (pdapp) transgenic mouse. J Neurosci 17: 7053–7059

Isaev NK, Stelmashook EV, Halle A, Harms C, Lautenschlager M, Weih M, Dirnagl U, Victorov IV, Zorov DB (2000) Inhibition of Na(+), K(+)-ATPase activity in cultured rat cerebellar granule cells prevents the onset of apoptosis induced by low potassium. Neurosci Lett 283: 41–4

Isgor C, Sengelaub DR (1998) Prenatal gonadal steroids affect adult spatial behavior, CA1 and CA3 pyramidal cell morphology in rats. Horm Behav 34: 183–98

Jaiswal RK, Weissinger E, Kolch W, Landreth GE (1996) Nerve growth factor-mediated activation of the mitogen-activated protein (MAP) kinase cascade involves a signaling complex containing B-Raf and HSP90. J Biol Chem 271: 23626–9

James VHT, Serio M, Giusti G (1976) The Endocrine Function of the Human Ovary. Proc Serono Symposia, Volume 7. Academic Press, London, New York, San Francisco

Jefferson WN, Newbold RR (2000) Potential endocrine-modulating effects of various phytoestrogens in the diet. Nutrition 16: 658–62

Jellinger KA (1999) The role of iron in neurodegeneration: prospects for pharmacotherapy of Parkinson's disease. Drug Aging 14: 115–40

Jensen EV, DeSombre ER (1973) Estrogen-receptor interaction. Science 182: 126–34

Jensen EV, Jacobson HI (1962) Basic guides to the mechanism of estrogen action. Recent Prog Horm Res 18: 387–414

Ji L, Mochon E, Arcinas M, Boxer LM (1996) CREB proteins function as positive regulators of the translocated bcl-2 allele in t(14–18) lymphomas. J Biol Chem 271: 22687–22691

Joel PB, Traish AM, Lannigan DA (1998) Estradiol-induced phosphorylation of serine 118 in the estrogen receptor is independent of p42/p44 mitogen-activated protein kinase. J Biol Chem 273: 13317–23

Joels M (1997) Steroid hormones and excitability in the mammalian brain. Front Neuroendocrin 18: 2–48

Jolly-Tornetta C, Wolf BA (2000) Regulation of amyloid precursor protein (APP) secretion by protein kinase C alpha in human ntera 2 neurons (NT2N). Biochemistry 39: 7428–7435

Jordan VC (1999) Targeted antiestrogens to prevent breast cancer. Trends Endocrin Met 10: 312–317

Jung-Testas I, Baulieu EE (1998) Steroid hormone receptors and steroid action in rat glial cells of the central and peripheral nervous system. J Steroid Biochem Molec Biol 65: 243–51

Kaipia A, Hsueh AJW (1997) Regulation of ovarian follicle atresia. Annu Rev Physiol 59: 349–363

Kaltschmidt B, Uherek M, Volk B, Baeuerle PA, Kaltschmidt C (1997) Transcription factor NF-kappaB is activated in primary neurons by amyloid beta peptides and in neurons surrounding early plaques from patients with Alzheimer disease. Proc Natl Acad Sci USA 94: 2642–7

Kaltschmidt C, Kaltschmidt B, Baeuerle PA (1993) Brain synapses contain inducible forms of the transcription factor NF-kappa B. Mech Dev 43: 135–47

Kaltschmidt C, Kaltschmidt B, Neumann H, Wekerle H, Baeuerle PA (1994) Constitutive NF-kappa B activity in neurons. Mol Cell Biol 14: 3981–92

Kampen DL, Sherwin BB (1994) Estrogen use and verbal memory in healthy postmenopausal women. Obstet Gynecol 83: 979–83

Kandel ES, Hay N (1999) The regulation and activities of the multifunctional serine/threonine kinase Akt/PKB. Exp Cell Res 253: 210–229

Karas RH, Patterson BL, Mendelsohn ME (1994) Human vascular smooth muscle cells contain functional estrogen receptor. Circulation 89: 1943–50

Karin M (1999) How NF-kappa B is activated: the role of the I kappa B kinase (IKK) complex. Oncogene 18: 6867–6874

Kato S, Endoh H, Masuhiro Y, Kitamoto T, Uchiyama S, Sasaki H, Masushige S, Gotoh Y, Nishida E, Kawashima H, et al (1995) Activation of the estrogen receptor through phosphorylation by mitogen-activated protein kinase. Science 270: 1491–4

Kato S, Masuhiro Y, Watanabe M, Kobayashi Y, Takeyama K, Endoh H, Yanagisawa J (2000) Molecular mechanism of a cross-talk between oestrogen and growth factor signalling pathways. Gen Cells 5: 593–601

Katzenellenbogen JA, Katzenellenbogen BS (1996) Nuclear hormone receptors: ligand-activated regulators of transcription and diverse cell responses. Chem Biol 3: 529–36

Kavaliers M, Ossenkopp KP, Prato FS, Innes DG, Galea LA, Kinsella DM, Perrot-Sinal TS (1996) Spatial learning in deer mice: sex differences and the effects of endogenous opioids and 60 Hz magnetic fields. J Comp Physiol A Neural Behav Physiol 179: 715–24

Kawakami M, Yoshioka E, Konda N, Arita J, Visessuvan S (1978) Data on the sites of stimulatory feedback action of gonadal steroids indispensable for luteinizing hormone release in the rat. Endocrinol 102: 791–8

Kawas C, Resnick S, Morrison A, Brookmeyer R, Corrada M, Zonderman A, Bacal C, Lingle DD, Metter E (1997) A prospective study of estrogen replacement therapy and the risk of developing Alzheimer's disease: the Baltimore Longitudinal Study of Aging. Neurology 48: 1517–21

Kelley ST, Thackray VG (1999) Phylogenetic analyses reveal ancient duplication of estrogen receptor isoforms. J Mol Evol 49: 609–14

Kew JN, Smith DW, Sofroniew MV (1996) Nerve growth factor withdrawal induces the apoptotic death of developing septal cholinergic neurons in vitro: protection by cyclic AMP analogue and high potassium. Neuroscience 70: 329–39

Kim-Schulze S, McGowan KA, Hubchak SC, Cid MC, Martin MB, Kleinman HK, Greene GL, Schnaper HW (1996) Expression of an estrogen receptor by human coronary artery and umbilical vein endothelial cells. Circulation 94: 1402–7

Kimura D (1992) Sex differences in the brain. Sci Am 267: 118–25

King WJ, Greene GL (1984) Monoclonal antibodies localize oestrogen receptor in the nuclei of target cells. Nature 307: 745–7

Klaiber EL, Broverman DM, Vogel W, Kobayashi Y (1979) Estrogen therapy for severe persistent depressions in women. Arch Gen Psychiatry 36(5): 550–4

Klaiber EL, Broverman DM, Vogel W, Peterson LG, Snyder MB (1996) Individual differences in changes in mood and platelet monoamine oxidase (MAO) activity during hormonal replacement therapy in menopausal women. Psychoneuroendocrinology 21: 575–92

Klein-Hitpass L, Ryffel GU, Heitlinger E, Cato AC (1988) A 13 bp palindrome is a functional estrogen responsive element and interacts specifically with estrogen receptor. Nucl Acids Res 16: 647–63

Klein-Hitpass L, Schwerk C, Kahmann S, Vassen L (1998) Targets of activated steroid hormone receptors: basal transcription factors and receptor interacting proteins. J Mol Med 76: 490–6

Kliewer SA, Lehmann JM, Willson TM (1999) Orphan nuclear receptors: shifting endocrinology into reverse. Science 284: 757–60

Klinge CM (2000) Estrogen receptor interaction with co-activators and co-repressors. Steroids 65: 227–51

Knazek RA, Lippmann ME, Chopra HC (1977) Formation of solid human mammary carcinoma in vitro. J Natl Cancer Inst 58: 419–22

Knierim JJ (2000) LTP takes route in the hippocampus. Neuron 25: 504–6

Kompoliti K (1999) Estrogen and movement disorders. Clin Neuropharmacol 22: 318–26

Konnecke R, Hafner H, Maurer K, Loffler W, an der Heiden W (2000) Main risk factors for schizophrenia: increased familial loading and pre- and peri-natal complications antagonize the protective effect of oestrogen in women. Schizophr Res 44: 81–93

Korach KS (1994) Insights from the study of animals lacking functional estrogen receptor. Science 266: 1524–7

Korach KS, Couse JF, Curtis SW, Washburn TF, Lindzey J, Kimbro KS, Eddy EM, Migliaccio S, Snedeker SM, Lubahn DB, Schomberg DW, Smith EP (1996) Estrogen receptor gene disruption: molecular characterization and experimental and clinical phenotypes. Recent Prog Horm Res 51: 159–88

Kow LM, Mobbs CV, Pfaff DW (1994) Roles of second-messenger systems and neuronal activity in the regulation of lordosis by neurotransmitters, neuropeptides, and estrogen: a review. Neurosci Biobehav R 18: 251–68

Krege JH, Hodgin JB, Couse JF, Enmark E, Warner M, Mahler JF, Sar M, Korach KS, Gustafsson JA, Smithies O (1998) Generation and reproductive phenotypes of mice lacking estrogen receptor beta. Proc Natl Acad Sci USA 95: 15677–82

Kudo T, Imaizumi K, Tanimukai H, Katayama T, Sato N, Nakamura Y, Tanaka T, Kashiwagi Y, Jinno Y, Tohyama M, Takeda M (2000) Are cerebrovascular factors involved in Alzheimer's disease? Neurobiol Aging 21: 215–224

Kueng W, Wirz-Justice A, Menzi R, Chappuis-Arndt E (1976) Regional brain variations of tryptophan, monoamines, monoamine oxidase activity, plasma free and total tryptophan during the estrous cycle of the rat. Neuroendocrinology 21: 289–96

Kuhn HG, Dickinson-Anson H, Gage FH (1996) Neurogenesis in the dentate gyrus of the adult rat: age-related decrease of neuronal progenitor proliferation. J Neurosci 16: 2027–33

Kuiper GG, Brinkmann AO (1994) Steroid hormone receptor phosphorylation: is there a physiological role? Mol Cell Endocrinol 100: 103–7

Kuiper GG, Enmark E, Pelto-Huikko M, Nilsson S, Gustafsson JA (1996) Cloning of a novel receptor expressed in rat prostate and ovary. Proc Natl Acad Sci USA 93: 5925–30

Kuiper GG, Shughrue PJ, Merchenthaler I, Gustafsson JA (1998) The estrogen receptor beta subtype: a novel mediator of estrogen action in neuroendocrine systems. Front Neuroendocrin 19: 253–86

Kuiper GG, van den Bemd GJ, van Leeuwen JP (1999) Estrogen receptor and the SERM concept. J Endocrinol Invest 22: 594–603

Kurzchalia TV, Parton RG (1999) Membrane microdomains and caveolae. Curr Opin Cell Biol 11: 424–31

Lacort M, Leal AM, Liza M, Martin C, Martinez R, Ruiz-Larrea MB (1995) Protective effect of estrogens and catecholestrogens against peroxidative membrane damage in vitro. Lip 30: 141–6

Lacroix AZ, Burke W (1997) Breast cancer and hormone replacement therapy. Lancet 350: 1042–1043

Lafferty FW, Fiske ME (1994) Postmenopausal estrogen replacement: a long-term cohort study. Am J Med 97: 66–77

Lambert JJ, Belelli D, Hill-Venning C, Peters JA (1995) Neurosteroids and GABAA receptor function. Trends Pharmacol Sci 16: 295–303

Landfield PW, Eldridge JC (1994) The glucocorticoid hypothesis of age-related hippocampal neurodegeneration: role of dysregulated intraneuronal calcium. Annals NY Acad Sci 746: 308–26

Landfield PW (1988) Hippocampal neurobiological mechanisms of age-related memory dysfunction. Neurobiol Aging 9: 571–9

Langub MC Jr, Watson RE Jr (1992) Estrogen receptor-immunoreactive glia, endothelia, and ependyma in guinea pig preoptic area and median eminence: electron microscopy. Endocrinology 130: 364–72

Lantin-Hermoso RL, Rosenfeld CR, Yuhanna IS, German Z, Chen Z, Shaul PW (1997) Estrogen acutely stimulates nitric oxide synthase activity in fetal pulmonary artery endothelium. Am J Physiol 273: L119–26

Le Mellay V, Grosse B, Lieberherr M (1997) Phospholipase C beta and membrane action of calcitriol and estradiol. J Biol Chem 272: 11902–7

Lee CK, Weindruch R, Prolla TA (2000) Gene-expression profile of the ageing brain in mice. Nat Genet 25: 294–7

Lee MM, Lin SS (2000) Dietary fat and breast cancer. Annu Rev Nutr 20: 221–248

Lee TH (2000) By the way, doctor... I'm 68, and I've been taking Viagra for about a year now. The drug is working for me, but I'm always a little scared that I am going to give myself a heart attack. Should I be? Harvard Health Letter 25: 8

Legler J, van den Brink CE, Brouwer A, Murk AJ, van der Saag PT, Vethaak AD, van der Burg B (1999) Development of a stably transfected estrogen receptor-mediated luciferase reporter gene assay in the human T47D breast cancer cell line. Toxicol Sci 48: 55–66

Lemmen JG, Broekhof JLM, Kuiper GGJM, Gustafsson JA, van der Saag PT, van der Burg B (1999) Expression of estrogen receptor alpha and beta during mouse embryogenesis. Mech Develop 81: 163–167

Leo C, Chen JD (2000) The SRC family of nuclear receptor coactivators. Gene 245: 1–11

Leroy E, Anastasopoulos D, Konitsiotis S, Lavedan C, Polymeropoulos MH (1998) Deletions in the Parkin gene and genetic heterogeneity in a Greek family with early onset Parkinson's disease. Hum Genet 103: 424–427

Levenson AS, Jordan VC (1998) The key to the antiestrogenic mechanism of raloxifene is amino acid 351 (aspartate) in the estrogen receptor. Cancer Res 58: 1872–1875

Levin ER (2000) Nuclear receptor versus plasma membrane oestrogen receptor. Novartis Foundation Symposium 230, 41–50

Lewin B (1974) Interaction of regulator proteins with recognition sequences of DNA. Cell 2: 1–7

Lewin B (2000) Genes VII. Oxford University Press, Cell Press, Cambridge

Lezoualc'h F, Behl C (1998) Transcription factor NF-κB: friend or foe of neurons? Mol Psychiatry 3: 15–2

Lezoualc'h F, Engert S, Berning B, Behl C (2000) Corticotropin-releasing hormone-mediated neuroprotection is associated with an increased release of non-amyloidogenic amyloid protein precursor and with the suppression of NF-κB. Mol Endocrinol 14: 147–159

Lezoualc'h F, Sparapani M, Behl C (1998) N-acetyl-serotonin (normelatonin) and melatonin protect neurons against oxidative challenges and suppress the activity of the transcription factor NF-κB. J Pin Res 24: 168–178

Lezoualc'h F, Rupprecht R, Holsboer F, Behl C (1996) Bcl-2 prevents hippocampal cell death induced by the neuroleptic drug haloperidol. Brain Res 738: 176–9

Li HY, Bian JS, Kwan YW, Wong TM (2000) Enhanced responses to 17beta-estradiol in rat hearts treated with isoproterenol: Involvement of a cyclic AMP-dependent pathway. J Pharmacol Exp Ther 293: 592–598

Li NX, Karin M (1999) Is NF-kappa B the sensor of oxidative stress? FASEB J 13: 1137–1143

Li X, Schwartz PE, Rissman EF (1997) Distribution of estrogen receptor-beta-like immunoreactivity in rat forebrain. Neuroendocrinology 66: 63–7

Liaw JJ, He JR, Hartman RD, Barraclough CA (1992) Changes in tyrosine hydroxylase mRNA levels in medullary A1 and A2 neurons and locus coeruleus following castration and estrogen replacement in rats. Brain Res Mol Brain Res 13: 231–8

Lien EJ, Ren S, Bui HH, Wang R (1999) Quantitative structure-activity relationship analysis of phenolic antioxidants. Free Radical Bio Med 26: 285–294

Lindner V, Kim SK, Karas RH, Kuiper GG, Gustafsson JA, Mendelsohn ME (1998) Increased expression of estrogen receptor-beta mRNA in male blood vessels after vascular injury. Circ Res 83: 224–9

Lipshutz RJ, Morris D, Chee M, Hubbell E, Kozal MJ, Shah N, Shen N, Yang R, Fodor SP (1995) Using oligonucleotide probe arrays to access genetic diversity. Biotechniques 19: 442–7

Lipton SA (1997) Janus faces of NF-κB: neurodestruction versus neuroprotection. Nature Med 3: 20–22

Liu GT, Zhang TM, Wang BE, Wang YW (1992) Protective action of seven natural phenolic compounds against peroxidative damage to biomembranes. Biochem Pharmacol 43: 147–152

Liu ZG, Hsu HL, Goeddel DV, Karin M (1996) Dissection of TNF receptor 1 effector functions: JNK activation is not linked to apoptosis while NF-κB activation prevents cell death. Cell 87: 565–576

Longcope C (1986) Adrenal and gonadal androgen secretion in normal females. Clin Endocrinol Metab 15: 213–28

Longcope C, Franz C, Morello C, Baker R, Johnston CC Jr (1986) Steroid and gonadotropin levels in women during the peri-menopausal years. Maturitas 8: 189–96

Loo DT, Copani A, Pike CJ, Whittemore ER, Walencewicz AJ, Cotman CW (1993) Apoptosis is induced by beta-amyloidin cultured central nervous system neurons. Proc Natl Acad Sci USA 90: 7951–5

Losordo DW, Kearney M, Kim EA, Jekanowski J, Isner JM (1994) Variable expression of the estrogen receptor in normal and atherosclerotic coronary arteries of premenopausal women. Circulation 89: 1501–10

Love RR (1995) Tamoxifen Chemoprevention – Public health goals, toxicities for all and benefits to a few. Ann Oncol 6: 127–128

Love RR, Mazess RB, Barden HS, Epstein S, Newcomb PA, Jordan VC, Carbone PP, DeMets DL (1992) Effects of tamoxifen on bone mineral density in postmenopausal women with breast cancer. N Engl J Med 326: 852–6

Love RR, Wiebe DA, Newcomb PA, Cameron L, Leventhal H, Jordan VC, Feyzi J, DeMets DL (1991) Effects of tamoxifen on cardiovascular risk factors in postmenopausal women. Ann Intern Med 115: 860–4

Love RR, Barden HS, Mazess RB, Epstein S, Chappell RJ (1994) Effect of tamoxifen on lumbar spine bone mineral density in postmenopausal women after 5 years. Arch Intern Med 154 (22): 2585–8

Love S (1999) Oxidative stress in brain ischemia. Brain Pathol 9: 119–131

Lubahn DB, Moyer JS, Golding TS, Couse JF, Korach KS, Smithies O (1993) Alteration of reproductive function but not prenatal sexual development after insertional disruption of the mouse estrogen receptor gene. Proc Natl Acad Sci USA 90: 11162–6

Lucassen PJ, Chung WC, Kamphorst W, Swaab DF (1997) DNA damage distribution in the human brain as shown by in situ end labeling; area-specific differences in aging and Alzheimer disease in the absence of apoptotic morphology. J Neuropathol Exp Neurol 56: 887–900

Lucking CB, Abbas N, Durr A, Bonifati V, Bonnet AM, Debroucker T, Demichele G, Wood NW, Agid Y, Brice A (1998) Homozygous deletions in parkin gene in european and north african families with autosomal recessive juvenile parkinsonism. Lancet 352: 1355–1356

Luine VN, Khylchevskaya RI, McEwen BS (1975) Effect of gonadal steroids on activities of monoamine oxidase and choline acetylase in rat brain. Brain Res 86: 293–306

Luine VN, Richards ST, Wu VY, Beck KD (1998) Estradiol enhances learning and memory in a spatial memory task and effects levels of monoaminergic neurotransmitters. Horm Behav 34: 149–62

Lupien SJ, Nair NP, Briere S, Maheu F, Tu MT, Lemay M, McEwen BS, Meaney MJ (1999)

coactivators: multiple enzymes, multiple complexes, multiple functions. J Steroid Biochem 69: 3–12

Melov S, Ravenscroft J, Malik S, Gill MS, Walker DW, Clayton PE, Wallace DC, Malfroy B, Doctrow SR, Lithgow GJ (2000) Extension of life-span with superoxide dismutase/catalase mimetics. Science 289: 1567–1569

Mendelsohn ME, Karas RH (1994) Estrogen and the blood vessel wall. Curr Opin Cardiol 9: 619–26

Mendelsohn ME, Karas RH (1999) The protective effects of estrogen on the cardiovascular system. N Engl J Med 340: 1801–11

Mendis T, Suchowersky O, Lang A, Gauthier S (1999) Management of Parkinson's disease – A review of current and new therapies. Can J Neurol Sci 26: 89–103

Menini A (1999) Calcium signalling and regulation in olfactory neurons. Curr Opin Neurobiol 9: 419–426

Menon KM, Gunaga KP (1974) Role of cyclic AMP in reproductive processes. Fertil Steril 25: 732–50

Merchenthaler I, Shugrue PJ (1999) Estrogen receptor-beta: a novel mediator of estrogen action in brain and reproductive tissues. Morphological considerations. J Endocrinol Invest 22 [10 Suppl]: 10–2

Mermelstein PG, Becker JB, Surmeier DJ (1996) Estradiol reduces calcium currents in rat neostriatal neurons via a membrane receptor. J Neurosci 16: 595–604

Merry DE, Korsmeyer SJ (1997) BCL-2 gene family in the nervous system. Annu Rev Neurosci 20: 245–267

Meyer M, Schreck R, Baeuerle PA (1993) H_2O_2 and antioxidants have opposite effects on activation of NF-kB and AP-1 in intact cells: AP-1 as secondary antioxidant-responsive factor. EMBO J 12: 2005–2015

Meyers MJ, Sun J, Carlson KE, Katzenellenbogen BS, Katzenellenbogen JA (1999) Estrogen receptor subtype-selective ligands: asymmetric synthesis and biological evaluation of cis- and trans-5,11-dialkyl- 5,6,11, 12-tetrahydrochrysenes. J Med Chem 42: 2456–68

Migliaccio A, Di Domenico M, Castoria G, de Falco A, Bontempo P, Nola E, Auricchio F (1996) Tyrosine kinase/p21ras/MAP-kinase pathway activation by estradiol-receptor complex in MCF-7 cells. EMBO J 15: 1292–300

Milgrom E, Atger M, Baulieu EE (1973) Studies on estrogen entry into uterine cells and on estradiol-receptor complex attachment to the nucleus – is the entry of estrogen into uterine cells a protein-mediated process? Biochim Biophys Act 320: 267–83

Miller DB, Ali SF, O'Callaghan JP, Laws SC (1998) The impact of gender and estrogen on striatal dopaminergic neurotoxicity. Ann NY Acad Sci 844: 153–65

Miller DK (1997) The role of the Caspase family of cysteine proteases in apoptosis. Semin Immunol 9: 35–49

Miller MM, Bennett HP, Billiar RB, Franklin KB, Joshi D (1998) Estrogen, the ovary, and neurotransmitters: factors associated with aging. Exp Gerontol 33: 729–57

Miller S, Mayford M (1999) Cellular and molecular mechanisms of memory: the LTP connection. Curr Opin Genet Develop 9: 333–7

Mills J, Laurent Charest D, Lam F, Beyreuther K, Ida N, Pelech SL, Reiner PB (1997) Regulation of amyloid precursor protein catabolism involves the mitogen-activated protein kinase signal transduction pathway. J Neurosci 17: 9415–22

Mogil JS (1999) The genetic mediation of individual differences in sensitivity to pain and its inhibition. Proc Natl Acad Sci USA 96: 7744–51

Mooradian AD (1993) Antioxidant properties of steroids. J Steroid Biochem Mol Biol 45: 509–11

Moosmann B, Behl C (2000) Cytoprotective antioxidant function of tyrosine and tryptophan residues in transmembrane proteins. Eur J Biochem 267: 5687–5692

Moosmann B, Behl C (1999) The antioxidant neuroprotective effects of estrogens and phenolic compounds are independent from their estrogenic properties. Proc Natl Acad Sci USA 96: 8867–8872

Moosmann B, Behl C (2000) Dietary phenols: Antioxidants for the brain? Nutr Neurosci 3: 1–10

Moosmann B, Uhr M, Behl C (1997) Neuroprotective potential of aromatic alcohols against oxidative cell death. FEBS Lett 413: 467–472

Mor G, Nilsen J, Horvath T, Bechmann I, Brown S, Garcia-Segura LM, Naftolin F (1999) Estrogen and microglia: A regulatory system that affects the brain. J Neurobiol 40: 484–96

Morey AK, Pedram A, Razandi M, Prins BA, Hu RM, Biesiada E, Levin ER (1997) Estrogen and progesterone inhibit vascular smooth muscle proliferation. Endocrinology 138: 3330–9

Moss RL, Gu Q, Wong M (1997) Estrogen: nontranscriptional signaling pathway. Recent Prog Horm Res 52: 33–69

Mucke L, Masliah E, Johnson WB, Ruppe MD, Alford M, Rockenstein EM, Forss-Petter S, Pietropaolo M, Mallory M, Abraham CR (1994) Synaptotropic effects of human amyloid beta protein precursors in the cortex of transgenic mice. Brain Res 666: 151–167

Mulnard RA, Cotman CW, Kawas C, van Dyck CH, Sano M, Doody R, Koss E, Pfeiffer E, Jin S, Gamst A, Grundman M, Thomas R, Thal LJ (2000) Estrogen replacement therapy for treatment of mild to moderate Alzheimer disease: a randomized controlled trial. Alzheimer's Disease Cooperative Study. JAMA 283: 1007–15

Munaro NI (1978) The effect of ovarian steroids on hypothalamic 5-hydroxytryptamine neuronal activity. Neuroendocrinology 26: 270–6

Murphy DD, Segal M (1997) Morphological plasticity of dendritic spines in central neurons is mediated by activation of cAMP response element binding protein. Proc Natl Acad Sci USA 94: 1482–7

Naftolin F, Ryan KJ (1975) The metabolism of androgens in central neuroendocrine tissues. J Steroid Biochem 6: 993–7

Nakai M, Qin ZH, Chen JF, Wang Y, Chase TN (2000) Kainic acid-induced apoptosis in rat striatum is associated with nuclear factor-kappaB activation. J Neurochem 74: 647–58

Nakano M, Sugioka K, Naito I, Takekoshi S, Niki E (1987) Novel and potent biological antioxidants on membrane phospholipid peroxidation: 2-hydroxyestrone and 2-hydroxy estradiol. Biochem Biophys Res Commun 142: 919–24

Namba H, Sokoloff L (1984) Acute administration of high doses of estrogen increases glucose utilization throughout brain. Brain Res 291: 391–4

Nardulli AM, Greene GL, Shapiro DJ (1993) Human estrogen receptor bound to an estrogen response element bends DNA. Mol Endocrinol 7: 331–40

Nass SJ, Davidson NE (1999) The biology of breast cancer. Hematol Oncol Clin N 13: 311

Newell-Price J, Trainer P, Besser M, Grossman A (1998) The diagnosis and differential diagnosis of Cushing's syndrome and pseudo-Cushing's states. Endocrin Rev 19: 647–72

Nicholson RI, Gee JMW (2000) Oestrogen and growth factor cross-talk and endocrine insensitivity and acquired resistance in breast cancer. Br J Cancer 82: 501–513

Noble BS, Reeve J (2000) Osteocyte function, osteocyte death and bone fracture resistance. Mol Cell Endocrinol 159: 7–13

Noble BS, Stevens H, Loveridge N, Reeve J (1997) Identification of apoptotic changes in osteocytes in normal and pathological human bone. Bone 20: 273–82

Olanow CW (1993) A radical hypothesis for neurodegeneration. Trends Neurosci 16: 439–444

Oldendorf WH (1971) Brain uptake of radiolabeled amino acids, amines, and hexoses after arterial injection. Am J Physiol 221: 1629–1639

Orwoll ES, Klein RF (1995) Osteoporosis in men. Endocr Rev 16: 87–116

Osborne CK, Fuqua SAW (2000) Selective estrogen receptor modulators: Structure, function, and clinical use. J Clin Oncol 18: 3172–3186

Osterlund M, Kuiper GG, Gustafsson JA, Hurd YL (1998) Differential distribution and regu-

lation of estrogen receptor-alpha and -beta mRNA within the female rat brain. Brain Res Mol Brain Res 54: 175–80
Oyama Y, Fuchs PA, Katayama N, Noda K (1994) Myricetin and quercetin, the flavonoid constituents of *Gingko biloba* extract, greatly reduce oxidative metabolism in both resting and Ca^{2+}-loaded brain neurons. Brain Res 635: 125–129
Packard MG, Teather LA (1997) Intra-hippocampal estradiol infusion enhances memory in ovariectomized rats. Neuroreport 8: 3009–13
Packard MG, Teather LA (1997) Posttraining estradiol injections enhance memory in ovariectomized rats: cholinergic blockade and synergism. Neurobiol Learn Mem 68: 172–88
Paech K, Webb P, Kuiper GG, Nilsson S, Gustafsson J, Kushner PJ, Scanlan TS (1997) Differential ligand activation of estrogen receptors ERalpha and ERbeta at AP1 sites. Science 277: 1508–10
Paganini-Hill A (1995) Estrogen replacement therapy and stroke. Prog Cardiovasc Dis 38: 223–42
Pahl HL, Baeuerle PA (1994) Oxygen and the control of gene expression. Bioessays 16: 497–502
Palumbo A, Yeh J (1995) Apoptosis as a basic mechanism in the ovarian cycle: follicular atresia and luteal regression. J Soc Gynecol Invest 2: 565–73
Pappas TC, Gametchu B, Watson CS (1995) Membrane estrogen receptors identified by multiple antibody labeling and impeded-ligand binding. FASEB J 9: 404–10
Pardridge WM, Moeller TL, Mietus LJ, Oldendorf WH (1980) Blood-brain barrier transport and brain sequestration of steroid hormones. Am J Physiol 239: E96–E102
Parkinson Study Group (1993) Effects of tocopherol and deprenyl on the progression of disability in early Parkinson's disease. N Engl J Med 328: 176–83
Parnham M, Sies H, Ebselen (2000) Prospective therapy for cerebral ischaemia. Expert Opin Invest Drugs 9: 607–619
Parrizas M, Saltiel AR, Leroith D (1997) Insulin-like growth factor 1 inhibits apoptosis using the phosphatidylinositol 3′-kinase and mitogen-activated protein kinase pathways. J Biol Chem 272: 154–161
Parthasarathy S, Santanam N, Ramachandran S, Meilhac O (1999) Oxidants and antioxidants in atherogenesis: an appraisal. J Lipid Res 40: 2143–2157
Parthasarathy S, Steinberg D, Witztum JL (1992) The role of oxidized low-density lipoproteins in the pathogenesis of atherosclerosis. Annu Rev Med 43: 219–225
Paul SM, Purdy RH (1992) Neuroactive steroids. FASEB J 6: 2311–22
Pearlstein T, Rosen K, Stone AB (1997) Mood disorders and menopause. Endocrin Metab Clin 26: 279–94
Pecins-Thompson M, Brown NA, Bethea CL (1998) Regulation of serotonin re-uptake transporter mRNA expression by ovarian steroids in rhesus macaques. Brain Res Mol Brain Res 53: 120–9
Pecins-Thompson M, Brown NA, Kohama SG, Bethea CL (1996) Ovarian steroid regulation of tryptophan hydroxylase mRNA expression in rhesus macaques. J Neurosci 16: 7021–9
Peschanski M, Defer G, N'Guyen JP, Ricolfi F, Monfort JC, Remy P, Geny C, Samson Y, Hantraye P, Jeny R et al (1994) Bilateral motor improvement and alteration of L-dopa effect in two patients with Parkinson's disease following intrastriatal transplantation of foetal ventral mesencephalon. Brain 117: 487–99
Petanceska SS, Nagy V, Frail D, Gandy S (2000) Ovariectomy and 17beta-estradiol modulate the levels of Alzheimer's amyloid beta peptides in brain. Neurology 54: 2212–2217
Pettersson K, Grandien K, Kuiper GG, Gustafsson JA (1997) Mouse estrogen receptor beta forms estrogen response element-binding heterodimers with estrogen receptor alpha. Mol Endocrinol 11: 1486–96
Pfaff DW (1997) Hormones, genes, and behavior. Proc Natl Acad Sci USA 94: 14213–6
Phillips SM, Sherwin BB (1992) Variations in memory function and sex steroid hormones across the menstrual cycle. Psychoneuroendocrinology 17: 497–506

Pietras RJ, Szego CM (1975) Endometrial cell calcium and oestrogen action. Nature 253: 357–9

Pietras RJ, Szego CM (1977) Specific binding sites for oestrogen at the outer surfaces of isolated endometrial cells. Nature 265: 69–72

Pietras RJ, Szego CM (1980) Partial purification and characterization of oestrogen receptors in subfractions of hepatocyte plasma membranes. Biochem J 191: 743–60

Pike AC, Brzozowski AM, Roberts SM, Olsen OH, Persson E (1999) Structure of human factor VIIa and its implications for the triggering of blood coagulation. Proc Natl Acad Sci USA 9: 8925–30

Pike CJ (1999) Estrogen modulates neuronal Bcl-xL expression and beta-amyloid-induced apoptosis: relevance to Alzheimer's disease. J Neurochem 72: 1552–63

Pike MC, Krailo MD, Henderson BE, Casagrande JT, Hoel DG (1983) 'Hormonal' risk factors, 'breast tissue age' and the age-incidence of breast cancer. Nature 303: 767–70

Platet N, Cunat S, Chalbos D, Rochefort H, Garcia M (2000) Unliganded and liganded estrogen receptors protect against cancer invasion via different mechanisms. Mol Endocrinol 14: 999–1009

Plouffe L (2000) Selective estrogen receptor modulators (SERMs) in clinical practice. J Soc Gynecol Invest 7: S38–S46

Poewe WH, Wenning GK (1998) The natural history of parkinsons-disease. Ann Neurol 44: S1–S9

Polkowski K, Mazurek AP (2000) Biological properties of genistein. A review of in vitro and in vivo data. Act Pol Pharm 57: 135–55

Polo-Kantola P, Portin R, Polo O, Helenius H, Irjala K, Erkkola R (1998) The effect of short-term estrogen replacement therapy on cognition: a randomized, double-blind, cross-over trial in postmenopausal women. Obstet Gynecol 91: 459–66

Polymeropoulos MH, Lavedan C, Leroy E, Ide SE, Dehejia A, Dutra A, Pike B, Root H, Rubenstein J, Boyer R, Stenroos ES, Chandrasekharappa S, Athanassiadou A, Papapetropoulos T, Johnson WG, Lazzarini AM, Duvoisin RC, Diiorio G, Golbe LI, Nussbaum RL (1997) Mutation in the alpha-synuclein gene identified in families with Parkinsons disease. Science 276: 2045–2047

Popovici RM, Kao LC, Giudice LC (2000) Discovery of new inducible genes in in vitro decidualized human endometrial stromal cells using microarray technology. Endocrinol 141: 3510–3

Porter M, Penney GC, Russell D, Russell E, Templeton A (1996) A population based survey of women's experience of the menopause. Br J Obstet Gynaecol 103: 1025–8

Post A, Holsboer F, Behl C (1998) Induction of nf-kappa-b activity during haloperidol-induced oxidative toxicity in clonal hippocampal cells – suppression of nf-kappa-b and neuroprotection by antioxidants. J Neurosci 18: 8236–8246

Prasad KN, Clarkson ED, Larosa FG, Edwardsprasad J, Freed CR (1998) Efficacy of grafted immortalized dopamine neurons in an animal model of parkinsonism – a review. Mol Genet Met 65: 1–9

Prasad KN, Cole WC, Kumar B (1999) Multiple antioxidants in the prevention and treatment of Parkinson's disease. J Am Coll Nutr 18: 413–423

Prasad KN, Hovland AR, Cole WC, Prasad KC, Nahreini P, Edwards-Prasad J, Andreatta CP (2000) Multiple antioxidants in the prevention and treatment of Alzheimer disease: Analysis of biologic rationale. Clin Neuropharmacol 23: 2–13

Priest CA, Pfaff DW (1995) Actions of sex steroids on behaviours beyond reproductive reflexes. Ciba Foundation Symposium 191: 74–89

Rachez C, Freedman LP (2000) Mechanisms of gene regulation by vitamin D(3) receptor: a network of coactivator interactions. Gene 246: 9–21

Raff MC, Barres BA, Burne JF, Coles HS, Ishizaki Y, Jacobson MD (1993) Programmed cell death and the control of cell survival: lessons from the nervous system. Science 262: 695–700

Ramirez VD, Zheng J (1996) Membrane sex-steroid receptors in the brain. Front Neuroendocrin 17: 402–39

Razandi M, Pedram A, Greene GL, Levin ER (1999) Cell membrane and nuclear estrogen receptors (ERs) originate from a single transcript: studies of ERalpha and ERbeta expressed in Chinese hamster ovary cells. Mol Endocrinol 13: 307–19

Redmond GP (1999) Hormones and sexual function. Intern J Fertil Wom Med 44: 193–7

Regier DA, Hirschfeld RM, Goodwin FK, Burke JD Jr, Lazar JB, Judd LL (1988) The NIMH Depression Awareness, Recognition, and Treatment Program: structure, aims, and scientific basis. Am J Psychiatry 145: 1351–7

Reiter RJ (1998) Oxidative damage in the central nervous system: protection by melatonin. Progr Neurobiol 56: 359–384

Resnick SM, Maki PM, Golski S, Kraut MA, Zonderman AB (1998) Effects of estrogen replacement therapy on PET cerebral blood flow and neuropsychological performance. Hormon Behav 34: 171–82

Revelli A, Massobrio M, Tesarik J (1998) Nongenomic actions of steroid hormones in reproductive tissues. Endocr Rev 19: 3–17

Ribeiro RCJ, Kushner PJ, Baxter JD (1995) The nuclear hormone receptor gene superfamily. Annu Rev Med 46: 443–453

Riddoch D, Jefferson M, Bickerstaff ER (1971) Chorea and the oral contraceptives. Brit Med J 4: 217–8

Rifici VA, Khachadurian AK (1992) The inhibition of low-density lipoprotein oxidation by 17-β estradiol. Metabolism 41: 1110–1114

Riggs BL, Melton LJ 3rd (1986) Involutional osteoporosis. N Engl J Med 314: 1676–86

Robinson D, Friedman L, Marcus R, Tinklenberg J, Yesavage J (1994) Estrogen replacement therapy and memory in older women. J Am Geriatr Soc 42: 919–22

Rochefort H, Platet N, Hayashido Y, Derocq D, Lucas A, Cunat S, Garcia M (1998) Estrogen receptor mediated inhibition of cancer cell invasion and motility – an overview. J Steroid Biochem Mol Biol 65: 163–168

Rodriguez J, Garcia de Boto MJ, Hidalgo A (1996) Mechanisms involved in the relaxant effect of estrogens on rat aorta strips. Life Sci 5: 607–15

Rogers J, Cooper NR, Webster S, Schultz J, McGeer PL, Styren SD, Civin WH, Brachova L, Bradt B, Ward P, et al (1992) Complement activation by b-amyloid in Alzheimer disease. Proc Natl Acad Sci USA 89, 10016–10020

Roof RL, Hall ED (2000) Title Gender differences in acute CNS trauma and stroke: Neuroprotective effects of estrogen and progesterone. J Neurotraum 17: 367–388

Roof RL, Havens MD (1992) Testosterone improves maze performance and induces development of a male hippocampus in females. Brain Res 572: 310–3

Roof RL, Zhang Q, Glasier MM, Stein DG (1993) Gender-specific impairment on Morris water maze task after entorhinal cortex lesion. Behav Brain Res 57: 47–51

Rosen DR, Siddique T, Patterson D, Figlewicz DA, Sapp P, Hentati A, Donaldson D, Goto J, O'Regan JP, Deng H-X, Rhamani Z, Krizus A, McKenna-Yasek D, Cayabyab A, Gaston SM, Berger R, Tanzi RE, Halperin JJ, Herzfeldt B, Vanden-Bergh R, Hung W-Y, Bird T, Deng G, Mulder DW, Smyth C, Laing NG, Soriano E, Pericak-Vance MA, Haines J, Rouleau GA, Gusella JS, Horvitz HR, Brown RH Jr (1993) Mutations in the Cu/Zn superoxide dismutase gene are associated with familial amyotrophic lateral sclerosis. Nature 362: 59–62

Rosenzweig MR, Leiman AL, Breedlove SM (1999) Biological Psychology, 2nd edn. Sinauer Associates Inc, Sunderland Publishers

Roses AD (1997) Apolipoprotein E, a gene with complex biological interactions in the aging brain. Neurobiol Dis 4: 170–185

Ross PD, Davis JW, Epstein RS, Wasnich RD (1991) Pre-existing fractures and bone mass predict vertebral fracture incidence in women. Ann Intern Med 114: 919–23

Rubinow DR, Schmidt PJ, Roca CA (1998) Estrogen-serotonin interactions: implications for affective regulation. Biol Psychiatry 44: 839–50

Rupprecht R (1997) The neuropsychopharmacological potential of neuroactive steroids. J Psychiatry Res 31: 297–314

Rupprecht R, Holsboer F (1999) Neuroactive steroids: mechanisms of action and neuropsychopharmacological perspectives. Trends Neurosci 22: 410–6

Rupprecht R, Reul JM, Trapp T, van Steensel B, Wetzel C, Damm K, Zieglgansberger W, Holsboer F (1993) Progesterone receptor-mediated effects of neuroactive steroids. Neuron 11: 523–30

Rydel RE, Greene LA (1988) cAMP analogs promote survival and neurite outgrowth in cultures of rat sympathetic and sensory neurons independently of nerve growth factor. Proc Natl Acad Sci USA 85: 1257–61

Sack MN, Rader DJ, Cannon RO 3rd (1994) Oestrogen and inhibition of oxidation of low-density lipoproteins in postmenopausal women. Lancet 343: 269–70

Sagara Y (1998) Induction of reactive oxygen species in neurons by haloperidol. J Neurochem 71: 1002–12

Saibil H (2000) Molecular chaperones: containers and surfaces for folding, stabilising or unfolding proteins. Curr Opin Struc Biol 10: 251–8

Salas E, Lopez MG, Villarroya M, Sanchez-Garcia P, De Pascual R, Dixon WR, Garcia AG (1994) Endothelium-independent relaxation by 17-alpha-estradiol of pig coronary arteries. Eur J Pharmacol 258: 47–55

Sano M, Ernesto C, Thomas RG, Klauber MR, Schafer K, Grundma M, Woodbury P, Growdon J, Cotman CW, Pfeiffer E et al (1997) A controlled trial of selegiline, alpha-tocopherol, or both as treatment for Alzheimer's disease. N Engl J Med 336: 1216–1222

Santagati S, Melcangi RC, Celotti F, Martini L, Maggi A (1994) Estrogen receptor is expressed in different types of glial cells in culture. J Neurochem 63: 2058–64

Santanam N, Shern-Brewer R, McClatchey R, Castellano PZ, Murphy AA, Voelkel S, Parthasarathy S (1998) Estradiol as an antioxidant: incompatible with its physiological concentrations and function. J Lipid Res 39: 2111–8

Sapolsky RM (1992) Stress, the aging brain, and the mechanisms of neuron death. MIT Press, Cambridge Massachusetts, London England

Sapolsky RM (1996) Why stress is bad for your brain. Science 273: 749–50

Sapolsky RM (1999) Glucocorticoids, stress, and their adverse neurological effects: relevance to aging. Exp Gerontol 34: 721–32

Saunders-Pullman R, Gordon-Elliott J, Parides M, Fahn S, Saunders HR, Bressman S (1999) The effect of estrogen replacement on early Parkinson's disease. Neurology 52: 1417–1421

Sawada H, Ibi M, Kihara T, Urushitani M, Akaike A, Shimohama S (1998) Estradiol protects mesencephalic dopaminergic neurons from oxidative stress-induced neuronal death. J Neurosci Res 54: 707–719

Sawada H, Ibi M, Kihara T, Urushitani M, Honda K, Nakanishi M, Akaike A, Shimohama S (2000) Mechanisms of antiapoptotic effects of estrogens in nigral dopaminergic neurons. FASEB J 14: 1202–1214

Sawada H, Shimohama S (2000) Neuroprotective effects of estradiol in mesencephalic dopaminergic neurons. Neurosci Biobehav Rev 24: 143–147

Scallet AC (1999) Estrogens: neuroprotective or neurotoxic? Ann NY Acad Sci 890: 121–32

Schena M, Shalon D, Heller R, Chai A, Brown PO, Davis RW (1996) Parallel human genome analysis: microarray-based expression monitoring of 1000 genes. Proc Natl Acad Sci USA 93: 10614–9

Schenk D, Barbour R, Dunn W, Gordon G, Grajeda H, Guido T, Hu K, Huang JP, Johnson-Wood K, Khan K, Kholodenko D, Lee M, Liao ZM, Lieberburg I, Motter R, Mutter L, Soriano F, Shopp G, Vasquez N, Vandevert C, Walker S, Wogulis M, Yednock T, Games D, Seubert P

(1999) Immunization with amyloid-beta attenuates Alzheimer disease-like pathology in the PDAPP mouse. Nature 400: 173–177

Schenk DB, Seubert P, Lieberburg I, Wallace J (2000) Beta-peptide immunization – A possible new treatment for Alzheimer disease. Arch Neurol 57: 934–936

Schmidt BMW, Gerdes D, Feuring M, Falkenstein E, Christ M, Wehling M (2000) Rapid, nongenomic steroid actions: A new age? Front Neuroendocrin 21: 57–94

Schneider A, Martin-Villalba A, Weih F, Vogel J, Wirth T, Schwaninger M (1999) NF-κB is activated and promotes cell death in focal cerebral ischemia. Nat Med 5 554–559

Schneider LS (1998) Cholinergic deficiency in Alzheimer's disease. Pathogenic model. Am J Geriatr Psych 6 [2 Suppl 1]: S49–55

Schneider LS, Farlow M (1997) Combined tacrine and estrogen replacement therapy in patients with Alzheimer's disease. Ann NY Acad Sci 826: 317–22

Schneider LS, Farlow MR, Pogoda JM (1997) Potential role for estrogen replacement in the treatment of Alzheimer's dementia. Am J Med 103: 46S–50S

Schneider LS, Finch CE (1997) Can estrogens prevent neurodegeneration? Drug Aging 11: 87–95

Schneider LS, Small GW, Hamilton SH, Bystritsky A, Nemeroff CB, Meyers BS (1997) Estrogen replacement and response to fluoxetine in a multicenter geriatric depression trial. Fluoxetine Collaborative Study Group. Am J Geriatr Psych 5: 97–106

Schreck R, Rieber P, Baeuerle PA (1991) Reactive oxygen intermediates as apparently widely used messengers in the activation of the NF–κB transcription factor and HIV-1. EMBO J 10: 2247–2258

Schubert D, Behl C (1993) The expression of amyloid beta protein precursor protects nerve cells from beta-amyloid and glutamate toxicity and alters their interaction with the extracellular matrix. Brain Res 629: 275–282

Schulz JB, Lindenau J, Seyfried J, Dichgans J (2000) Glutathione, oxidative stress and neurodegeneration. Eur J Biochem 267: 4904–4911

Seelig A, Gottschlich R, Devant RM (1994) A method to determine the ability of drugs to diffuse through the blood-brain barrier. Proc Natl Acad Sci USA 91: 68–72

Seeman MV (1996) The role of estrogen in schizophrenia. J Psychiatry Neurosci 21: 123–127

Seeman MV (1997) Psychopathology in women and men – focus on female hormones. Am J Psychiatry 154: 1641–1647

Selkoe DJ (1999) Translating cell biology into therapeutic advances in Alzheimer's disease. Nature 399: A23–31

Sen R, Baltimore D (1986) Multiple nuclear factors interact with the immunoglobulin enhancer sequences. Cell 46: 705–716

Sgonc R, Boeck G, Dietrich H, Gruber J, Recheis H, Wick G (1994) Simultaneous determination of cell surface antigens and apoptosis. Trends Genet 10: 41–2

Shaul PW (1999) Rapid activation of endothelial nitric oxide synthase by estrogen. Steroids 64: 28–34

Sherwin BB (1988) Estrogen and/or androgen replacement therapy and cognitive functioning in surgically menopausal women. Psychoneuroendocrinology 13: 345–57

Sherwin BB (2000) Mild cognitive impairment: Potential pharmacological treatment options. J Am Geriat Soc 48: 431–441

Shi J, Simpkins JW (1997) 17beta-Estradiol modulation of glucose transporter 1 expression in blood-brain barrier. Am J Physiol 272: E1016–22

Shiau AK, Barstad D, Loria PM, Cheng L, Kushner PJ, Agard DA, Greene GL (1998) The structural basis of estrogen receptor/coactivator recognition and the antagonism of this interaction by tamoxifen. Cell 95: 927–937

Shibata H, Spencer TE, Onate SA, Jenster G, Tsai SY, Tsai MJ, O'Malley BW (1997) Role of coactivators and co-repressors in the mechanism of steroid/thyroid receptor action. Recent Prog Horm Res 52: 141–65

Shoulson I (1998) Datatop – a decade of neuroprotective inquiry. Ann Neurol 44: S 160–S 166

Shughrue PJ, Lane MV, Merchenthaler I (1997) Comparative distribution of estrogen receptor-alpha and -beta mRNA in the rat central nervous system. J Compar Neurol 388: 507–25

Shughrue PJ, Merchenthaler I (2000) Estrogen is more than just a "sex hormone": novel sites for estrogen action in the hippocampus and cerebral cortex. Front Neuroendocrin 21: 95–101

Shughrue PJ, Scrimo PJ, Merchenthaler I (1998) Evidence for the colocalization of estrogen receptor-beta mRNA and estrogen receptor-alpha immunoreactivity in neurons of the rat forebrain. Endocrinology 139: 5267–70

Shughrue PJ, Scrimo PJ, Merchenthaler I (2000) Estrogen binding and estrogen receptor characterization (ERalpha and ERbeta) in the cholinergic neurons of the rat basal forebrain. Neuroscience 96: 41–9

Shumaker SA, Reboussin BA, Espeland MA, Rapp SR, McBee WL, Dailey M, Bowen D, Terrell T, Jones BN (1998) The Women's Health Initiative Memory Study (WHIMS): a trial of the effect of estrogen therapy in preventing and slowing the progression of dementia. Control Clin Trials 19: 604–21

Shwaery GT, Vita JA, Keaney JF Jr (1997) Antioxidant protection of LDL by physiological concentrations of 17beta-estradiol. Requirement for estradiol modification. Circulation 95: 1378–85

Shy H, Malaiyandi L, Timiras PS (2000) Protective action of 17beta-estradiol and tamoxifen on glutamate toxicity in glial cells. Int J Develop Neurosci 18: 289–97

Sies H (1993) Strategies of antioxidant defense. Eur J Biochem 215: 213–9

Sies H (1997) Oxidative stress: oxidants and antioxidants. Exp Physiol 82: 291–5

Sies H, Stahl W (1995) Vitamins E and C, beta-carotene, and other carotenoids as antioxidants. Am J Clin Nutr 62 [6 Suppl]: 1315S–1321S

Simoncini T, Hafezi-Moghadam A, Brazil DP, Ley K, Chin WW, Liao JK (2000) Interaction of oestrogen receptor with the regulatory subunit of phosphatidylinositol-3-OH kinase. Nature 407: 538–41

Simpkins JW, Green PS, Gridley KE, Singh M, de Fiebre NC, Rajakumar G (1997) Role of estrogen replacement therapy in memory enhancement and the prevention of neuronal loss associated with Alzheimer's disease. Am J Med 103: 19S–25S

Simpkins JW, Singh M, Bishop J (1994) The potential role for estrogen replacement therapy in the treatment of the cognitive decline and neurodegeneration associated with Alzheimer's disease. Neurobiol Aging 15 [Suppl 2]: S 195–7

Singer CA, Figueroa-Masot XA, Batchelor RH, Dorsa DM (1999) The mitogen-activated protein kinase pathway mediates estrogen neuroprotection after glutamate toxicity in primary cortical neurons. J Neurosci 19: 2455–2463

Singer CA, Rogers KL, Dorsa DM (1998) Modulation of Bcl-2 expression: a potential component of estrogen protection in NT2 neurons. Neuroreport 9: 2565–8

Singer TP, Ramsay, PR (1995) Monoamine oxidases: old friends hold many surprises. FASEB J 9: 605–610

Singh M, Meyer EM, Simpkins JW (1995) The effect of ovariectomy and estradiol replacement on brain-derived neurotrophic factor messenger ribonucleic acid expression in cortical and hippocampal brain regions of female Sprague-Dawley rats. Endocrinology 136: 2320–4

Singh M, Setalo G Jr, Guan X, Frail DE, Toran-Allerand CD (2000) Estrogen-induced activation of the mitogen-activated protein kinase cascade in the cerebral cortex of estrogen receptor-alpha knock-out mice. J Neurosci 20: 1694–1700

Singh M, Setalo G Jr, Guan X, Warren M, Toran-Allerand CD (1999) Estrogen-induced activation of mitogen-activated protein kinase in cerebral cortical explants: convergence of estrogen and neurotrophin signaling pathways. J Neurosci 19: 1179–88

Siuciak JA, Lewis DR, Wiegand SJ, Lindsay RM (1997) Antidepressant-like effect of brain-derived neurotrophic factor (BDNF). Pharmacol Biochem Behav 56: 131–7

Skaper SD, Fabris M, Ferrari V, Dalle Carbonare M, Leon A (1997) Quercetin protects cutaneous tissue-associated cell types including sensory neurons from oxidative stress induced by glutathione depletion: cooperative effects of ascorbic acid. Free Radical Bio Med 22: 669–678

Skoog I (2000) Vascular aspects in Alzheimer's disease. J Neural Transm [Suppl] (59): 37–43

Slemenda CW (1993) Risk factors for low bone mass: clinical implications. Ann Intern Med 118: 741–2

Smith MA, Rottkamp CA, Nunomura A, Raina AK, Perry G (2000) Oxidative stress in Alzheimer's disease. Biochim Biophys Act – Molecular Basis of Disease 1502: 139–144

Soderling TR (2000) CaM-kinases: modulators of synaptic plasticity. Curr Opin Neurobiol 10: 375–380

Sohrabji F, Miranda RC, Toran-Allerand CD (1994) Estrogen differentially regulates estrogen and nerve growth factor receptor mRNAs in adult sensory neurons. J Neurosci 14: 459–71

Sohrabji F, Miranda RC, Toran-Allerand CD (1995) Identification of a putative estrogen response element in the gene encoding brain-derived neurotrophic factor. Proc Natl Acad Sci USA 92: 11110–4

Spillantini MG, Goedert M (1998) Tau protein pathology in neurodegenerative diseases. Trends Neurosci 21: 428–433

Squire LR (1992) Memory and the hippocampus: a synthesis from findings with rats, monkeys, and humans. Psychol Rev 99: 195–231

Stampfer MJ, Colditz GA, Willett WC, Manson JE, Rosner B, Speizer FE, Hennekens CH (1991) Postmenopausal estrogen therapy and cardiovascular disease. Ten-year follow-up from the nurses' health study. N Engl J Med 325: 756–62

Steinberg D (1997) Low density lipoprotein oxidation and its pathobiological significance. J Biol Chem 272: 20963–6

Stone DJ, Rozovsky I, Morgan TE, Anderson CP, Finch CE (1998) Increased synaptic sprouting in response to estrogen via an apolipoprotein E-dependent mechanism: implications for Alzheimer's disease. J Neurosci 18: 3180–5

Stone DJ, Rozovsky I, Morgan TE, Anderson CP, Hajian H, Finch CE (1997) Astrocytes and microglia respond to estrogen with increased apoE mRNA in vivo and in vitro. Exp Neurol 143: 313–8

Strijks E, Kremer JA, Horstink MW (1999) Effects of female sex steroids on Parkinson's disease in postmenopausal women. Clin Neuropharmacol 22: 93–7

Studd J (1997) Depression and the menopause. Oestrogens improve symptoms in some middle aged women. BMJ 314: 977–8

Subbiah MTR, Kessel B, Agrawal M, Rajan R, Abplanalp W, Rymaszewski (1993) Antioxidant potential of specific estrogens and lipid peroxidation. J Clin Endocrinol Metab 77: 1095–1097

Sugioka K, Shimosegawa Y, Nakano M (1987) Estrogens as natural antioxidants of membrane phospholipid peroxidation. FEBS Lett 210: 37–9

Tan SL, Wood M, Maher P (1998) Oxidative stress induces a form of programmed cell death with characteristics of both apoptosis and necrosis in neuronal cells. J Neurochem 71: 95–105

Tanapat P, Hastings NB, Reeves AJ, Gould E (1999) Estrogen stimulates a transient increase in the number of new neurons in the dentate gyrus of the adult female rat. J Neurosci 19: 5792–801

Tang MX, Jacobs D, Stern Y, Marder K, Schofield P, Gurland B, Andrews H, Mayeux R (1996) Effect of oestrogen during menopause on risk and age of onset of Alzheimer's disease. Lancet 348: 429–432

Tao Y, Black IB, DiCicco-Bloom E (1996) Neurogenesis in neonatal rat brain is regulated by peripheral injection of basic fibroblast growth factor (bFGF). J Compar Neurol 376: 653–63

Tenover JL (1999) Testosterone replacement therapy in older adult men. Int J Androl 22: 300–6

Terao J, Piskula M, Yao Q (1994) Protective effect of epicatechin, epicatechin gallate, and quercetin on lipid peroxidation in phospholipid bilayers. Arch Biochem Biophys 308: 278–284

Terry RD, Katzman R, Bick KL (1994) Alzheimer Disease. Raven Press, New York

Tesarik J, Mendoza C (1995) Nongenomic effects of 17beta-estradiol on maturing human oocytes: relationship to oocyte developmental potential. J Clin Endocrinol Metab 80: 1438–43

Teyler TJ, Vardaris RM, Lewis D, Rawitch AB (1980) Gonadal steroids: effects on excitability of hippocampal pyramidal cells. Science 209: 1017–8

Thoenen H (1995) Neurotrophins and neuronal plasticity. Science 270: 593–8

Thompson EW, Paik S, Brunner N, Sommers CL, Zugmaier G, Clarke R, Shima TB, Torri J, Donahue S, Lippman ME et al (1992) Association of increased basement membrane invasiveness with absence of estrogen receptor and expression of vimentin in human breast cancer cell lines. J Cell Physiol 150: 534–44

Thorlacius S, Struewing JP, Hartge P, Olafsdottir GH, Sigvaldason H, Tryggvadottir L, Wacholder S, Tulinius H, Eyfjord JE (1998) Population-based study of risk of breast cancer in carriers of brca2 mutation. Lancet 352: 1337–1339

Thornberry NA, Lazebnik Y (1998) Caspases: enemies within. Science 281: 1312–6

Toft D, Gorski J (1966) A receptor molecule for estrogens: isolation from the rat uterus and preliminary characterization. Proc Natl Acad Sci USA 55: 1574–81

Tohgi H, Utsugisawa K, Yamagata M, Yoshimura M (1995) Effects of age on messenger RNA expression of glucocorticoid, thyroid hormone, androgen, and estrogen receptors in postmortem human hippocampus. Brain Res 700: 245–53

Tomkinson A, Reeve J, Shaw RW, Noble BS (1997) The death of osteocytes via apoptosis accompanies estrogen withdrawal in human bone. J Clin Endocrinol Metab 82: 3128–35

Toran-Allerand CD (1996) The estrogen/neurotrophin connection during neural development: is co-localization of estrogen receptors with the neurotrophins and their receptors biologically relevant? Dev Neurosci 18: 36–48

Toran-Allerand CD, Miranda RC, Bentham WD, Sohrabji F, Brown TJ, Hochberg RB, MacLusky NJ (1992) Estrogen receptors colocalize with low-affinity nerve growth factor receptors in cholinergic neurons of the basal forebrain. Proc Natl Acad Sci USA 89: 4668–72

Toran-Allerand CD, Singh M, Setalo G Jr (1999) Novel mechanisms of estrogen action in the brain: new players in an old story. Front Neuroendocrin 20: 97–121

Toung TJ, Traystman RJ, Hurn PD (1998) Estrogen-mediated neuroprotection after experimental stroke in male rats. Stroke 29: 1666–70

Traber MG, Sies H (1996) Vitamin E in humans: demand and delivery. Annu Rev Nutr 16: 321–47

Tremblay GB, Tremblay A, Copeland NG, Gilbert DJ, Jenkins NA, Labrie F, Giguere V (1997) Cloning, chromosomal localization, and functional analysis of the murine estrogen receptor beta. Mol Endocrinol 11: 353–65

Tremblay GB, Tremblay A, Labrie F, Giguere V (1999) Dominant activity of activation function 1 (AF-1) and differential stoichiometric requirements for AF-1 and -2 in the estrogen receptor alpha-beta heterodimeric complex. Mol Cell Biol 19: 1919–1927

Trent JM, Bittner M, Zhang J, Wiltshire R, Ray M, Su Y, Gracia E, Meltzer P, De Risi J, Penland L, Brown P (1997) Use of microgenomic technology for analysis of alterations in DNA copy number and gene expression in malignant melanoma. Clin Exp Immunol 107 [Suppl 1]: 33–40

Tsujimoto Y, Shimizu S (2000) Bcl-2 family: Life-or-death switch. FEBS Letters 466: 6–10

Van Antwerp DJ, Martin SJ, Kafri T, Green DR, Verma IM (1996) Suppression of TNF-κ-induced apoptosis by NF-κB. Science 274: 787–789

Vassar R, Bennett BD, Babu-Khan S, Kahn S, Mendiaz EA, Denis P, Teplow DB, Ross S, Amarante P, Loeloff R, Luo Y, Fisher S, Fuller L, Edenson S, Lile J, Jarosinski MA, Biere AL, Curran E, Burgess T, Louis JC, Collins F, Treanor J, Rogers G, Citron M (1999) Beta-secretase cleavage of Alzheimer's amyloid precursor protein by the transmembrane aspartic protease BACE. Science 286: 735–741

Veld BAI, Launer LJ, Hoes AW, Ott A, Hofman A, Breteler MMB, Stricker BHC, (1998) NSAIDs and incident Alzheimer's disease. The Rotterdam study. Neurobiol Aging 19: 607–611

Venkov CD, Rankin AB, Vaughan DE (1996) Identification of authentic estrogen receptor in cultured endothelial cells. A potential mechanism for steroid hormone regulation of endothelial function. Circulation 94: 727–33

Verheul HAM, Coelingh-Bennink HJT, Kenemans P, Atsma WJ, Burger CW, Eden JA, Hammar M, Marsden J, Purdie DW (2000) Effects of estrogens and hormone replacement therapy on breast cancer risk and on efficacy of breast cancer therapies. Maturitas 36: 1–17

Verma IM, Stevenson JK, Schwarz EM, Vanantwerp D, Miyamoto S (1995) REL/NF-KAPPA-B/I-KAPPA-B family – intimate tales of association and dissociation. Gen Dev 9: 2723–2735

Voet D, Voet JG, Pratt CW (1999) Fundamentals of Biochemistry. Jon Wiley & Sons, Inc.

Wagner GC, Tekirian TL, Cheo CT (1993) Sexual differences in sensitivity to methamphetamine toxicity. J Neural Transm 93: 67–70

Wagner JP, Black IB, DiCicco-Bloom E (1999) Stimulation of neonatal and adult brain neurogenesis by subcutaneous injection of basic fibroblast growth factor. J Neurosci 19: 6006–16

Walter P, Green S, Greene G, Krust A, Bornert JM, Jeltsch JM, Staub A, Jensen E, Scrace G, Waterfield M et al (1985) Cloning of the human estrogen receptor cDNA. Proc Natl Acad Sci USA 82: 7889–93

Walton M, Sirimanne E, Williams C, Gluckman P, Dragunow M (1996) The role of the cyclic AMP-responsive element binding protein (CREB) in hypoxic-ischemic brain damage and repair. Brain Res Mol Brain Res 43: 21–9

Walton MR, Dragunow M (2000) Is CREB a key to neuronal survival? Trends Neurosci 23: 48–53

Wang CY, Mayo MW, Baldwin AS Jr (1996) TNF- and cancer therapy-induced apoptosis: Potentiation by inhibition of NF-κB. Science 274: 784–787

Wang M, Seippel L, Purdy RH, Backstrom T (1996) Relationship between symptom severity and steroid variation in women with premenstrual syndrome: study on serum pregnenolone, pregnenolone sulfate, 5 alpha-pregnane-3,20-dione and 3 alpha-hydroxy-5 alpha-pregnan-20-one. J Clin Endocrinol Metab 81: 1076–82

Wang Q, Santizo R, Baughman VL, Pelligrino DA (1999) Estrogen provides neuroprotection in transient forebrain ischemia through perfusion-independent mechanisms in rats. Stroke 30: 630–636

Warmuth MA, Sutton LM, Winer EP (1997) A review of hereditary breast cancer – from screening to risk factor modification. Am J Med 102: 407–415

Warner M, Nilsson S, Gustafsson JA (1999) The estrogen receptor family. Curr Opin Obstet Gyn 11: 249–54

Watson SJ, Akil H (1999) Gene chips and arrays revealed: a primer on their power and their uses. Biol Psychiatry 45: 533–43

Watters JJ, Campbell JS, Cunningham MJ, Krebs EG, Dorsa DM (1997) Rapid membrane effects of steroids in neuroblastoma cells: effects of estrogen on mitogen activated protein kinase signalling cascade and c-fos immediate early gene transcription. Endocrinology 138: 4030–3

Watters JJ, Dorsa DM (1998) Transcriptional effects of estrogen on neuronal neurotensin gene expression involve cAMP/protein kinase A-dependent signaling mechanisms. J Neurosci 18: 6672–80

Weaver CE, Parkchung M, Gibbs TT, Farb, DH (1997) 17-beta-estradiol protects against NMDA-induced excitotoxicity by direct inhibition of NMDA receptors. Brain Res 761: 338–341

Weiner CP, Lizasoain I, Baylis SA, Knowles RG, Charles IG, Moncada S (1994) Induction of

calcium-dependent nitric oxide synthases by sex hormones. Proc Natl Acad Sci USA 91: 5212–6
Weinmaster G (1997) The ins and outs of notch signaling. Mol Cell Neurosci 9: 91–102
Wetzel CH, Hermann B, Behl C, Pestel E, Rammes G, Zieglgansberger W, Holsboer F, Rupprecht R (1998) Functional antagonism of gonadal steroids at the 5-hydroxytryptamine type 3 receptor. Mol Endocrinol 12: 1441–51
Weyler W, Hsu, YPP, Breakefield XO (1990) Biochemistry and genetics of monoaminooxidase. Pharmacol Ther 47: 391–417
White R, Lees JA, Needham M, Ham J, Parker M (1987) Structural organization and expression of the mouse estrogen receptor. Mol Endocrinol 1: 735–44
Whitehouse PJ, Price DL, Struble RG, Clark AW, Coyle JT, Delon MR (1982) Alzheimer's disease and senile dementia: loss of neurons in the basal forebrain. Science 215: 1237–9
Whitmarsh AJ, Davis RJ (1996) Transcription factor AP-1 regulation by mitogen-activated protein kinase signal transduction pathways. J Mol Med 74: 589–607
Whooley MA, Grady D, Cauley JA (2000) Postmenopausal estrogen therapy and depressive symptoms in older women. J Gen Int Med 15: 535–541
Willingham MC (1999) Cytochemical methods for the detection of apoptosis. J Histochem Cytochem 47: 1101–10
Wolf OT, Kudielka BM, Hellhammer DH, Torber S, McEwen BS, Kirschbaum C (1999) Two weeks of transdermal estradiol treatment in postmenopausal elderly women and its effect on memory and mood: verbal memory changes are associated with the treatment induced estradiol levels. Psychoneuroendocrinology 24: 727–741
Wolozin B, Behl C (2000) Mechanisms of neurodegeneration disorders part 1: Protein aggregates. Arch Neurol 57: 793–796
Wolozin B, Behl C (2000) Mechanisms of neurodegenerative disorders part 2: Control of cell death. Arch Neurol 57: 801–804
Wolozin B, Iwasaki K, Vito P, Ganjei K, Lacana E, Sunderland T, Zhao B, Kusiak J, Wasco W, D'Adamio L (1996) Participation of presenilin-2 in apoptosis: Enhanced basal activity conferred by Alzheimer mutation. Science 274: 1710–3
Wong M, Moss RL (1992) Long-term and short-term electrophysiological effects of estrogen on the synaptic properties of hippocampal CA1 neurons. J Neurosci 12: 3217–25
Woolley CS (1998) Estrogen-mediated structural and functional synaptic plasticity in the female rat hippocampus. Horm Behav 34: 140–8
Woolley CS (1999) Electrophysiological and cellular effects of estrogen on neuronal function. Crit Rev Neurobiol 13: 1–20
Woolley CS, McEwen BS (1992) Estradiol mediates fluctuation in hippocampal synapse density during the estrous cycle in the adult rat. J Neurosci 12: 2549–54
Woolley CS, McEwen BS (1993) Roles of estradiol and progesterone in regulation of hippocampal dendritic spine density during the estrous cycle in the rat. J Comp Neurol 336: 293–306
Wooten MW (1999) Function for NF-κB in neuronal survival: Regulation by atypical protein kinase C. J Neurosci Res 58: 607–611
Wyllie AH, Kerr JF, Currie AR (1980) Cell death: the significance of apoptosis. Int Rev Cytol 68: 251–306
Xiao L, Becker JB (1998) Effects of estrogen agonists on amphetamine-stimulated striatal dopamine release. Synapse 29: 379–91
Xu H, Gouras GK, Greenfield JP, Vincent B, Naslund J, Mazzarelli L, Fried G, Jovanovic JN, Seeger M, Relkin NR, Liao F, Checler F, Buxbaum JD, Chait BT, Thinakaran G, Sisodia SS, Wang R, Greengard P, Gandy S (1998) Estrogen reduces neuronal generation of Alzheimer beta-amyloid peptides. Nat Med 4: 447–451
Yaffe K, Sawaya G, Lieberburg I, Grady D (1998) Estrogen therapy in postmenopausal women: effects on cognitive function and dementia. JAMA 279: 688–95

Yagi K (1997) Female hormones act as natural antioxidants – a survey of our research. Acta Biochim Pol 44: 701–709

Yang D, Kuan C, Whitmarsh A, Rincon M, Zheng T, Davis R, Rakic P, Flavell R (1996) Decreased apoptosis in the brain and premature lethality in CPP32-deficient mice. Nature 389: 368–72

Yang F, Sun X, Beech W, Teter B, Wu S, Sigel J, Vinters HV, Frautschy SA, Cole GM (1998) Antibody to caspase-cleaved actin detects apoptosis in differentiated neuroblastoma and plaque-associated neurons and microglia in Alzheimer's disease. Am J Pathol 152: 379–89

Yang SH, Shi J, Day AL, Simpkins JW (2000) Estradiol exerts neuroprotective effects when administered after ischemic insult. Stroke 31: 745–749

Yankner BA (1996) Mechanisms of neuronal degeneration in Alzheimers disease. Neuron 16: 921–932

Young RA (2000) Biomedical discovery with DNA arrays. Cell 102: 9–15

Zhang Y, Dawson VL, Dawson TM (2000) Oxidative stress and genetics in the pathogenesis of Parkinson's disease. Neurobiol Dis 7: 240–250

Zhong LT, Sarafian T, Kane DJ, Charles AC, Mah SP, Edwards RH, Bredesen DE (1993) BCL-2 inhibits death of central neural cells induced by multiple agents. Proc Natl Acad Sci USA 90: 4533–7

Zhou Y, Watters JJ, Dorsa DM (1996) Estrogen rapidly induces the phosphorylation of the cAMP response element binding protein in rat brain. Endocrinology 137: 2163–6

Subject index

Aβ 63, 158
Aβ-precursor protein (APP) 63
Acetylcholin-Deficiency-Hypothesis 71
Acetylcholine 69, 82
acetylcholinesterase (ACh-esterase) 69
ACh 69
action potential 77
Activation Function-1 (AF-1) 36
Activation Function-2 (AF-2) 37
adenocorticotropin 12
adenylate cyclase 90
αERKO 9, 65
aging 99
agonist 8
Akt 167
Alois Alzheimer 116
Alzheimer's Disease 14, 55
amine hormones 3
amyloid β protein (Aβ) 63
amyloid-cascade-hypothesis 116
androgens 12
androstenedione 16, 17
antagonists 8
antidepressant drugs 73
antiestrogens 29
antioxidant 54
antioxidants 93
apolipoprotein E 118
ApoE4 118
apoptosis 96, 144
apoptotic bodies 146
APP 63, 118
arginine vasopressin 12
aromatase 75
aromatase-reaction 16
arteriosclerosis 13
ascorbate (vitamin C) 57
α-secretase 117
astresia 96
Astrocytes (astroglia) 74
ATG 43
atherosclerosis 53
α-tocopherol (vitamin E) 57, 175
autocrine 9
autoradiography 31

bad 162
bax 162
BDNF 85, 162
βERKO 9, 65
bcl-2 162
bcl-X_l 162
bcl-X_s 162
blood-brain-barrier (B) 60
brain-derived neurotrophic factor (BDNF) 134
breast cancer 13
β-secretase 117
β-secretase-enzyme (BACE) 165

CA1 78
CA3 78
cAMP 11
cAMP binding protein (CBP) 169
cannabinoids 178
caspases 147
catalase 57
cataracts, skin damage, arteriosclerosis to neurodegenerative caveolae 88
caveolin 89
catechol estrogens 23
CB 1 32
CBP 3 44
cDNA libraries 30
cerebral autosomal dominant arteriopathy with subcortical infarcts and leucoencephalopathy (CADASIL) 128
cerebral cortex 33
chain-breaking antioxidant 60
chinese hamster ovary (CHO) 50
CHO 50
cholesterol 16

choline acetyltransferase (ChAT) 69, 71
cholinergic neurons 69, 70
classical estrogen activity 12
climacterium 96
co-activators 27
combinatorial chemistry 39
conception 21
conjugated estrogens 138
contraception 13
co-repressors 27
corticotrophin-releasing-hormone (CRH) 28, 155
CREB 28
CRH-R1 155
cushing's syndrome 12
cyclic adenosine monophospate 11
cyclic AMP response element binding protein 28
cytochrome P450 aromatase 24

dehydroepiandrosterone 17
dementia 95
dentate gyrus (DG) 33, 78
Depression 73, 132
DEVD 148
DEVD-fmk 148
diabetes mellitus 13
diet 59
diet, flavonoids 59
dimerizatin 26
disorders 55
DNA-arrays 99, 183
DNA-binding domain 27
DNA-binding domain DBD 36
DNA-binding domain LBD 36
DNA fragmentation 149
dopamine 74
DRIPs 44

ebselen 129
EC_{50} 58
E2 17
endocrine 9
endocrine glands 2
entorhinal cortex 78
ESTs 183
17β-estradiol 16, 17
estrogen replacement therapy 13
estrogen response elements 26
estrone 17
estrophilin 6
estrous cycle 13
ERα 8
ERβ 8
EREs 26
ERK 50, 86
ERK1 86
ERK2 86
ERR 31
ER-related receptor (ERR) type 1 31
ER-related receptor (ERR) type 2 31
excitotoxicity 58
expression maps 9
expression profile 183
Expression Profiling 99
extracelluar-kinase (ERK) 50

Fenton-Reaction 55, 92, 175
flavonoids 61, 178
fluoexitin 73
free radicals 54, 91
functional genomics 9, 100, 182

$GABA_A$ 81
GABA (γ-amino-butyric acid) 81
GABA-receptor 81
genistein 40
Genomics 182
glucocorticoid receptors 40
glucose transporter-1 75
glutamate receptor 82
glutathione 57
glutathione peroxidase (GSH-Px) 57
glycine 82
gonadotropins 21
granule cells 78
γ-secretase 117
GSH-Px 57

haloperidol 158
heat shock proteins 22
HeLa 30
heterodimers 26
High-Throughput-Assays 160, 185
high throughput screening methods 39
hilus structure 79
hippocampus 19, 77
H_2O_2 54
Hoechst 33342 157
homodimers 26
hormone 2, 12
hormone replacement therapy (HRT) 96
hsp 70 26

hsp 90 26
5-HT_3-receptor 25
5-HT_3 (serotonine) 82
Human Genom Project (HGP) 182
hydrogen peroxide (H_2O_2) 54
hydroxyl radicals ($OH^{\cdot}$) 54
17α-hydroxyprogesterone 17
hypothalamus-pituitary-adrenal (HPA) axis
105
Hypothalamus-Pituitary-Ovary-Uterus-Axis
18

ICI 29
IκB 170
immediate early genes 41
immunohistochemistry 33
inflammation 99
in-situ-hybridization 32
Insulin-like-growth-factor-1 (IGF-1) 75, 167

JNK 86

knock-out mouse 63

LDH 157
L-dopa 61
learning 69
L-glutamate 158
ligand-binding domain 27
ligand-binding domain, LBD 37
lipid peroxidation 59
long-term potentiation (LTP) 77
low-density-lipoprotein 58
low-density-lipoprotein (LDL) 102

major depression 132
MAPK 50, 86, 163
MCF-7 30
membrane receptors 49
membrane steroid receptors 49
memory 69, 106
menarche 96
menopause 13, 21, 94
menstruation 21
mild cognitive impairment (MCI) 122
mineralocorticoids 12
mitogen-activated protein (MAP) kinase 86
mitogen-activated-protein-kinase (MAPK)
50
modulatory domain 27
mossy fibers 78
mRNAs 30
(MTT) 156
Multiple Sclerosis 75

necrosis 144
neuroactive steroid 49, 80
neuroactive steroids 94
neurodegeneration 95
neurohormone 2
neuroprotection 95
neurosteroids 80
NF-κB 90, 168
NGF 69
nitric oxide (NO) 53
nitric oxide synthase (NOS) 53
NLS, nuclear localization sequences 37
Northern Blotting 31
NO 53, 102
NOS 53
nuclear factor kB 57

$O_2^{\cdot-}$ 54
$OH^{\cdot}$ 54
ologodendrocytes 75
oocyte 21
osteoporosis 13, 103
ovariectomy 72
oxidative phosphorylation 92
oxidative stress 55, 94, 160
oxidative stress hypothesis 113
oxytocin 12

p38-MAPKs 86
palindrome 41
paracrine 9
Parkinson's Disease 55, 74
PCR 30, 65
phosphatidyl-inositol-3 (PI-3) kinase
54, 167
PI-3 54
polymerase chain reaction 30
polyphenols 59
post-partum psychosis 13
pregnedolone 17
presenilin 1 (PS1) 118, 152
presenilin 2 (PS2) 118, 152
progesterone 16, 17
progesterone receptors 50
programmed cell death, apoptosis 104
programmed cell death (PCD) 145
progynon 6
Propidium iodide 157
protein hormones 3

protein kinase A (PKA) 90
Proteomics 182
puberty 21
pyramidal cells 78

quercetin 59

Raloxifen 44, 48
RAPs 44
reactive oxygen species (ROS) 54
RIPs 44
RNA polymerase II 26
ROS 54, 91, 175

sAPP 87, 164
Scatchard-Plot 29
schizophrenia 130
Schwann cells 75
SDS-Page 35
selective estrogen receptor modulators 8
senile plaques 111
SERMs 47, 137
Serotonin (5-hydroxytryptamine) 72
sex behavior 13
sex differences 68
sex differentiation 13
sex maturation 13
SOD 57
Southern blotting 65
SRC-1 36, 44
SRC3 44
stem cells 186
steroid hormones 4
steroid receptors 28
steroid receptor co-activator-1 (SRC-1) 44
stress hormones glucocorticoids 12
superoxide ($O_2^{\cdot-}$) 54
superoxide dismutase (SOD) 57
synaptic plasticity 75

tamoxifen 29, 44
tau protein 111
testosterone 12, 16, 17
TF 43
transgenic mouse 63
2,4,6-trimethylphenol (TMP) 180
trimethylphenol (TMP) 102
tryptophan hydroxylase 73
TUNEL-labeling 150
tyrosine hydroxylase 74

vasopressin 28

Western Blotting 33, 34
Woman's Health Initiative Memory Study (WHIMS) 122

Springer-Verlag and the Environment

WE AT SPRINGER-VERLAG FIRMLY BELIEVE THAT AN international science publisher has a special obligation to the environment, and our corporate policies consistently reflect this conviction.

WE ALSO EXPECT OUR BUSINESS PARTNERS – PRINTERS, paper mills, packaging manufacturers, etc. – to commit themselves to using environmentally friendly materials and production processes.

THE PAPER IN THIS BOOK IS MADE FROM NO-CHLORINE pulp and is acid free, in conformance with international standards for paper permanency.